Nighthawks and Black Widows

13th Air Force Night Fighters in the South and Southwest Pacific

1943-1945

Terry M. Mays

Schiffer Military History
Atglen, PA

DEDICATION

For Al "Super Luke" Lukas
and all 13th Air Force veterans of the 419th NFS,
Detachment B 6th NFS, and 550th NFS

Book Design by Stephanie Daugherty.

Library of Congress Control Number: 2009932315

Printed in China.
ISBN: 978-0-7643-3344-6
We are interested in hearing from authors with book ideas on related topics.

Published by Schiffer Publishing Ltd.
4880 Lower Valley Road
Atglen, PA 19310
Phone: (610) 593-1777
FAX: (610) 593-2002
E-mail: Info@schifferbooks.com.
Visit our web site at: www.schifferbooks.com
Please write for a free catalog.
This book may be purchased from the publisher.
Please include $5.00 postage.
Try your bookstore first.

In Europe, Schiffer books are distributed by:
Bushwood Books
6 Marksbury Avenue
Kew Gardens
Surrey TW9 4JF, England
Phone: 44 (0) 20 8392-8585
FAX: 44 (0) 20 8392-9876
E-mail: Info@bushwoodbooks.co.uk.
Visit our website at: www.bushwoodbooks.co.uk

Contents

ACKNOWLEDGMENTS

I am indebted to many individuals for assistance while preparing this book. I trust that those whom I inadvertently forget to mention will forgive me. Naturally, this book is a product of not only considerable research at the Air University Library and Air Force Historical Research Agency, but also the collective memories of many Night Fighter veterans whom I had the pleasure of chatting with at squadron reunions and in separate conversations over the years. This book began many years ago as I sat night after night listening to the stories of Alphonse (Al) "Super Luke" Lukas, the "Chicago Yank" who joined the Royal Air Force and transferred his European-learned night fighter skills to the 419th Night Fighter Squadron. I did not realize then that I would have multiple opportunities to hear the adventures of so many Night Fighter air and ground crew members from his squadron, including Don Dessert, Dick Stewart, Jim Allen, Ralph Dunavant, Phil Rand, Edwin Fulghum, Bob O'Brien, John Radzieta, Harold Young, Sandy Bauman, Bill McConnell, Richard Pitman, Clifford Johnson, Eddy Ballard, Cleve Wilborn, Bob Fowble, James Girdis, John Graham, Clem Green, Carl Reubin, Bob Wright, Wayne Frederickson, Eddie Barkhouse, and Chuck Pomeroy, as well as many of their wives. MAJ Don Fuqua (U.S. Army), my S3 Operations officer while deployed to the Middle East and a budding academic in his own right, reviewed the draft manuscript for clarity. Thanks to Jeff Kolln for support in acquiring some of the photographs in this book. Photos and maps were provided by many night fighter veterans as well as the U.S. Army, the Franklin Delano Roosevelt Presidential Library, and the Air Force Historical Research Agency (AFHRA). Jane Buskirk and Betsey Carter of The Citadel's Daniel library were instrumental in acquiring whatever published resource was needed for this project. Last, but certainly not least, I must thank Leslee Lukas Mays, my loving wife, best friend, and the daughter of "Super Luke" Lukas—that former Royal Air Force pilot who added so much personality to what would become the 419th Night Fighter Squadron of the 13th Air Force.

1

Formation and Royal Air Force Origins of the 419th NFS

"The sooner you [Americans] put [night] intruder operations into effect against the Japanese, the sooner the war in the Pacific will be won."

Flight Lieutenant Karel M. Kuttelwascher
Czech Night Fighter Ace, Royal Air Force

"Night fighting is an applied science that demands intense training."

Captain Al Lukas
pilot, 419th NFS

A history of the 13th Air Force Night Fighters originates not in the Pacific, but in the skies over Europe. One must look to Great Britain, where American pilots were learning the art of night fighter tactics while flying with the Royal Air Force (RAF) and Royal Canadian Air Force (RCAF) between 1940 and 1942. By the time the United States entered World War II Americans had developed their own bomber and day fighter training programs, but they learned night fighting from the British and Canadians. Two complete night fighter squadrons and one detachment of a third squadron flew with the 13th Air Force in the South and Southwest Pacific between 1943 and 1945. This book is their story, and one point of origin begins over Holland in the spring of 1942.[1]

The moonlight shined peacefully on May 6, 1942, as the RCAF Boston III (American lend-lease Douglas A-20C), nicknamed *Yankee Doodle* for its all-American crew, banked and headed back toward the German airfield at Gilze, Holland. Pilot Officer Alphonse "Al" Lukas, a future pilot with the 419th Night Fighter Squadron (NFS) of the 13th Air Force, strained as he attempted to spot any activity at the German fighter base.

Suddenly, the ground lights along the runway flicked on, indicating the presence of returning German night fighters. Lukas guided his bomber to the far end of the base and scanned the sky with his exceptional night vision. Within seconds, the landing lights of returning Junkers JU88s snapped on as the Germans began their quick landing sequence. Other RAF and RCAF pilots had attempted to creep behind landing German planes and machine gun them. However, this tactic rarely worked. As soon as the Germans saw the British tracers they would extinguish their landing lights and disappear into the night.

Lukas arrived over Gilze with his own way of doing things. As a JU88 maneuvered to land at the airfield,

An "All American" crew flying an American manufactured Boston bomber, appropriately nicknamed *Yankee Doodle* for the Royal Canadian Air Force's No. 418 Squadron. Left to right: Sergeant Bill Randolph (navigator from Memphis), Pilot Officer Al Lukas (pilot from Chicago), and Sergeant Harry Haskell (gunner from Boston). In 1942, Lukas personally scored this night fighter squadron's first aerial victory by downing a German Ju-88 through the unorthodox tactic of dropping his bomb load on the enemy aircraft in mid-air. In 1944 Lukas scored the first 13th Air Force Northrop P-61 Black Widow kill of a Japanese aircraft. *Courtesy of Leslee Lukas Mays.*

Lukas skillfully guided his plane to a position just behind and above the German aircraft. Lukas opened his bomb bay door to reveal three 250lb high explosive impact and ninety 4lb incendiary eggs nestled inside the plane's belly. When the German pilot reached the end of the runway, Lukas throttled his engines and flew over the JU88 as he released his bomb load. Rear gunner Sergeant Harry Haskell, of Boston, reported seeing a ball of flame as the JU88 crashed onto the runway while the remaining bombs rained down upon the hangars and a gasoline storage area.

Searchlights added to the illumination of the night sky as the Germans attempted to locate the intruder over their airfield. Lukas barreled across the field at an altitude of 50 feet in order to escape from the scene. The RCAF pilot swept across the Dutch countryside, leapfrogging windmills in a dash for the English Channel and the safety of England. Sergeant Haskell remembered that the fires at the Gilze airfield were still visible after flying sixty miles. The trio, which included Sergeant Bill Randolph, the navigator from Memphis, returned to England from their sixth combat mission as the first aircrew of RCAF Number 418 Squadron to score an aerial victory. Rather than sleeping upon their return, Lukas, Randolph, and Haskell sat in the Bradwell Bay mess facility all morning waiting for a reply to the after action report they filed upon landing. A message finally arrived by late morning noting that Dutch resistance operatives had confirmed that the German aircraft was destroyed before it could land. Lukas not only knocked the landing JU88 out of the sky, but also damaged five other JU88s on the ground![2]

The RAF officially awarded Lukas an aerial victory as a medium bomber pilot. Lukas proudly added a swastika underneath his cockpit window. This unconventional aerial victory, utilizing his bomb load, proved to be more satisfying to the pilot than the downing of a Japanese Mitsubishi Ki-21 Sally bomber with the 419th NFS of the 13th Air Force two years later. American night fighters flying with the RAF and RCAF like Lukas would form the backbone of America's own nocturnal fighter units and training program less than 10 months after the aerial engagement near Gilze.

The story of the 13th Air Force Night Fighters is primarily the history of the 419th NFS, which served as the organization's sole night fighter squadron during most of its existence in World War II. However, it is also the story of Detachment B 6th NFS (also known as Detachment B, Night Fighter Squadron, between September and December 1943) and the 550th NFS. Detachment B 6th NFS arrived with five aircraft in March 1943 and provided night fighter coverage in the 13th Air Force area of operations until November 1943, when some of the personnel departed and others were absorbed into the newly arrived 419th NFS, which would remain with the organization for the duration of the war. In January 1945, the 550th NFS officially became part of the 13th Air Force and served until the conclusion of the war. With the 13th Air Force night fighters built primarily around the 419th NFS, it is appropriate to open with an examination of this unit's formation, which also illustrates the development of the American night fighter program.

The story of the 419th NFS is not simply a narration of aircraft; it's a story based on the unique union of men and machines. The two components are inseparable. The 419th NFS employed the Douglas P-70 Nighthawk in the intruder role in the Solomon Islands; the Lockheed P-38 Lightning for aerial intercepts and intruding in the Solomon Islands and along northern New Guinea; and the Northrop P-61 Black Widow for aircraft interception and intruding in the New Guinea area and in the Philippines. While these were incredible machines, it was the men on the ground and the men in the cockpits who placed them into action against the Japanese. A unit history of the 419th NFS would not be complete without their stories. Half of the original 15 pilots of the 419th NFS were ex-RAF and RCAF night fighter veterans, while the other half were rookies straight out of American flying schools. The ground personnel included kids who had recently graduated from high school as well as college students and graduates.

A review of 13th Air Force operations is an opportunity to discover the incredible feats of aerial bravery and brotherly camaraderie of Americans at war. Lukas (419th NFS) and his crew, utilizing RCAF tactics while

Royal Air Force (RAF) Douglas Boston Bomber. The RAF utilized these American bombers as night fighters for intruder raids against German targets in occupied Europe. The RAF developed night fighter tactics in World War I. Many American pilots flew with RAF and Royal Canadian Air Force (RCAF) night fighter units and transferred their skills to the American Army Air Force beginning in 1942. Half of the original pilots of the American 419th Night Fighter Squadron deployed to the 13th Air Force with RAF or RCAF combat experience. *Courtesy of Leslee Lukas Mays.*

The Douglas Boston bombers utilized by the Royal Air Force and Royal Canadian Air Force as night fighters were variants of the Douglas A-20, seen here being moved from a factory hangar in the United States. *Courtesy of the Franklin Delano Roosevelt Library.*

flying a P-70, completed the first solo American night intruder raid of the Japanese headquarters at Rabaul. First Lieutenant Harold Michels (419th NFS) and his crew bagged a Japanese Kawasaki Ki-45 Nick fighter over Borneo—after flying a P-61 all the way from the southern Philippines. First Lieutenant Richard "Dick" Stewart (419th NFS) successfully tackled a Japanese Mitsubishi G4M Betty bomber while racing toward it on a collision course in a P-38 over Middleburg Island. Captain Earl Bennett (Detachment B 6th NFS) and his crew made the first kill of a Japanese aircraft with the underpowered P-70 Nighthawk during the war. These incidents only hint at what these men accomplished. It is also a story of hardship and loss. The 419th NFS lost twenty-eight aircraft due to training/operational accidents and combat between November 1943 and August 1945. Twenty-five aircrew members died in these incidents. Detachment B 6th NFS lost at least five more aircraft, and the 550th NFS lost at least two. The aircrews of the night fighter squadrons accepted the challenge of their profession without reservation and complaint, and should be recognized as some of the greatest unsung heroes of the United States Army Air Force (USAAF) during World War II.

The United States entered World War II without an effective night fighter force, although there had been plans since 1940 to design and develop an aircraft capable of intercepting enemy bombers at night. The P-61 Black Widow, the first American aircraft designed specifically for night fighting, would emerge from those designs and become combat operational in 1944. In the meantime, the USAAF faced a two-part dilemma. First, the United States lacked a night fighter that could fill the gap until the arrival of the P-61. Second, she lacked experienced crews ready to train an American night fighter force to meet the threats posed in a two-front war.

The introduction of the Douglas P-70 met the first dilemma. The RAF flew modified Douglas A-20 Havocs as night intruders, and these served as the models for an aircraft that would become known as the P-70 Nighthawk. The United States altered the A-20 from the original night fighter version flown by Al Lukas and other RAF and RCAF pilots. The British version of the A-20, a modified DB-7 Havoc, served as a medium bomber with a crew of three men including a pilot, navigator, and a rear gunner. The British added cannon to a later version to provide additional firepower for the night intruder aircrews patrolling over the European continent. The USAAF modified the design by placing four 20-mm cannon in a ventral tub located in the bomb bay area, painting the Plexiglas of the forward cockpit, and adding a British AI Mk IV radar system in the nose, as well as reducing the crew by eliminating the gunner. A radar operator replaced the navigator and sat in a cockpit behind the pilot. Later versions included the addition of machine guns in lieu of the ventral tub cannons and introduction of the new American SCR-540 radar in the nose.

Lieutenant Colonel Winston "Winkie" Kratz helped solve the second dilemma. In mid-1942, the USAAF dispatched Kratz to Great Britain to seek Americans serving in the RAF or RCAF night fighter squadrons. Kratz met Lukas and many other Americans to discuss night fighter tactics practiced by the British and Canadians. When he returned to the United States in February 1943, 25 young men accompanied him to Florida. These RAF and RCAF veterans, now holding USAAF commissions, would form the backbone of America's own night fighter program and several would serve in the 13th Air Force.[3]

The USAAF established its new night fighter school at Orlando and Kissimmee, Florida. The 348th Night Fighter Training Squadron conducted Phase 1 of the night fighter school in Orlando. Pilots received

First Lieutenant Al Lukas and other former Royal Air Force and Royal Canadian Air Force American pilots agreed to help develop and train American night fighter units. Lukas stopped in Dakar, Senegal, on his journey to Florida, where the United States established its first night fighter training bases. *Courtesy of Leslee Lukas Mays.*

Far left is Al Lukas. Far right is Major Bill Odell, and to his left is Czech night fighter ace Flight Lieutenant Karel Kuttelwascher (drinking beer). Odell participated in the first American raid over occupied Europe on July 4, 1942, and would later command the 547th Night Fighter Squadron. Kuttelwascher visited Florida to share night fighter tactics with the American crews undergoing training. *Courtesy of Leslee Lukas Mays.*

Standing in front of a Northrop P-70 Nighthawk. Left to right: First Lieutenant Al Lukas, Brigadier General Earl Barnes, Brigadier General Winston Kratz, Flight Lieutenant Karel Kuttelwascher (a Czech officer), Unknown, General Kubeta (a Czech officer), Unknown, and Major Bill Odell. *Courtesy of Leslee Lukas Mays.*

introductory training in night flying while the Radar Observers (R/Os) attended a separate program. Upon graduation, pilot-R/O teams were formed for Phase 2 training in Kissimmee. The 349th Night Fighter Training Squadron oversaw this phase, which was based on the team concept to allow the pilots and R/Os to become comfortable working together. The teams then moved to Dunnellon, Florida, for the final phase of their training, which prepared them for assignments to squadrons deploying overseas. The ground personnel received their own training in the maintenance of the P-70 aircraft and the SCR 540 radar system and then joined the pilot-R/O teams to form an operational squadron. The USAAF activated the 419th NFS on April 1, 1943, and trained its personnel between April and October 1943.

The USAAF brought many Allied specialists to Florida to observe the night fighter training and provide advice on this particular skill. In the summer of 1943, Flight Lieutenant Karel Kuttelwascher, a Czech night fighter ace flying with the RAF, arrived in Florida. Kuttelwascher escaped to France when the German army rolled into his native Czechoslovakia. He offered his services to the French Air Force and destroyed seven bombers during the German offensive through the Low Countries into France. Initially after the fall of France, Kuttelwascher remained as a pilot with the pro-German Vichy French Air Force. However, when the Vichy government ordered an air strike against a British naval task force shelling the French fleet harbored at Oran, in North Africa, Kuttelwascher and several other pilots refused to attack the British and escaped to England. Kuttelwascher joined the RAF and downed three German Me-109 fighters while flying Hurricanes for the British. In May 1941 the RAF converted his squadron into a night fighter unit and the Czech pilot scored another 16 aerial victories while flying night missions in Hurricanes.

Kuttelwascher specialized in intruder operations with the RAF. Intruders lurked over the European continent at night seeking targets of opportunity—orbiting German aircraft or ground targets, such as trains. Kuttelwascher brought his special knowledge of intruder counter measures to the fledgling American aviators training in Florida. During his two years of night missions, Kuttelwascher had witnessed nearly every German counter measure to mask targets and entrap RAF intruder pilots. The Germans would construct fake airfields to attract the attention of intruder pilots and waste their ordnance. Kuttelwascher recalled:

"They had three dummy fields at St. Andre, but I served with the French Air Force and I knew every field in that sector. They'd light up the dummy fields, and keep the real field as dim as possible. Generally, the dummy field trick was a failure."

Another favorite German deception involved baiting intruders with an easy airborne target. Kuttelwascher noted:

"The 'bait bomber' was fairly effective. I recall one night swooping low over a German field and finding a slow-moving German bomber in front of me. I came in for the kill, but had a feeling that something was wrong. The Nazi was moving too slowly. I realized at the last moment that he was bait and that German night fighters were above me and would come in for the kill as soon as I came in range. It was too late to stop, so I came in fast,

Royal Canadian Air Force No. 418 Squadron crews receive their final afternoon briefings before commencing night patrols over the European continent. Each plane patrolled a specific area for a set length of time looking for German targets of opportunity. Crews became very savvy at recognizing ruses to lure them into areas patrolled by German night fighters or concealed gun positions on the ground. *Courtesy of Leslee Lukas Mays.*

Pilot Officer Al Lukas transferred to the American Army Air Force in 1942 as a First Lieutenant. He arrived with credit for 19 combat missions over occupied Europe, during which he downed one German aircraft, damaged five planes on the ground, destroyed seven trains, and damaged one flak vessel. Note the "USA" tab on the shoulder of his Royal Air Force uniform. *Courtesy of Leslee Lukas Mays.*

destroyed the bait bomber, and sped out of danger before the Nazi night fighters closed in. From that time on, we avoided the slow-moving bombers and concentrated on planes that came in at normal speed."

The Germans also utilized a related tactic. Rather than flying a bait aircraft, they would construct a device that resembled an aircraft from the air at night and drive it along a runway to simulate a taxiing aircraft. Kuttelwascher recalled:

"For a while we did a lot of strafing of taxiing aircraft, but later avoided that type of operation unless we were certain of our targets. The Germans had a trick of fixing up a gun on a mobile platform and dressing the entire mount to look like a plane. They'd light it and run it up and down the field, and from the air it looked exactly like a taxiing plane. When the Hurricanes would dive on the 'plane' the gun would open up. It was fairly lethal at first."[4]

The RAF and RCAF also employed deception in the night fighter war over the continent of Europe, and Lukas, as well as other Americans, brought these lessons and tactics to the men of the 419th NFS training in Florida. The most important rule of intruding involved never flying a path more than once; always keep your foes guessing as to where and when you will strike. The 419th NFS would later lose a P-70 and its crew between Bougainville and Rabaul when they violated this rule. A second form of deception practiced in Europe was based on knowing your enemy's communication codes. British intelligence often knew *Luftwaffe* night signals and utilized this to help their intruders penetrate German air space. On September 7, 1942, Lukas began his eighteenth RCAF combat mission with a ruse. As his new Boston bomber, the *Yankee Doodle II,* crossed the French coast, German anti-aircraft batteries detected and opened up on the crew. The plane's gunner fired a single red flare, which was the German signal for that evening indicating that one of their returning night fighters was lost and attempting to locate its home airfield. The German gunners bought the ruse and ceased firing on Lukas and his crew. Finding no activity at their assigned target area, Lukas and his crew opted to destroy two trains before calling it a night.

The USAAF night fighter training curriculum mirrored the types of operations flown by RAF and RCAF pilots, including intruder operations. On the European continent, intruder operations included bombing airfields and other fixed targets, as well as strafing trains and other vehicles. The former RAF and RCAF pilots involved in the USAAF night fighter training program helped develop a series of scenarios to teach these intruder tactics to the fledgling pilots in Florida.

Practicing the art of destroying trains emerged as an interesting element in the American night fighter training program. Again, the Americans based their training on the lessons learned from the RAF and RCAF in their night fighter campaign against the Germans. In 1942, RCAF Number 418 Squadron proved very successful at intercepting and destroying trains in occupied Europe, and Lukas exhibited a special knack for the skill. The Germans prohibited most civilians from traveling in trains late at night, and this form of curfew helped minimize any non-German casualties among rail passengers after dark. While the Germans may not have planned the restriction as a protection for civilians, the curfew certainly worked to that end. The RCAF officially credited Lukas with the destruction of seven trains—twice accounting for more than one in a single night of intruding over the European continent.

Lukas tried his hand at train busting during his sixteenth RCAF mission on August 27, 1942. British intelligence knew the exact movements of German trains, including departure times and routes. RAF and RCAF night fighter units often planned their intruder missions around these schedules. The new Havocs flown by Number 418 Squadron at this time were armed with cannon—a natural for intercepting German trains as they raced along rail lines unable to dodge an intruder pilot. Lukas located a train that night by spotting its dim blackout lights and initiated a dive to intercept it. As the Havoc roared over the train Lukas peppered it with his forward firing cannon. The rail engineers halted the train after Lukas completed his first pass and uncoupled the box cars before continuing along the track in the engine. Lukas swept in again and fired his four cannons, only to see the train engine ignite, explode, and then flip over. He then completed several passes along the line of box cars while his gunner blasted away at them with the twin 30 caliber machine guns. The crew later spotted a second train that night and destroyed it before heading back to their roost in England. Mechanics later found cinders in the plane's engine from the train's smokestack, indicating how low Lukas flew when engaging the second locomotive. On September 4, 1942, Lukas topped this total when he and his crew destroyed three trains in France during a single mission.

Lukas and others developed ingenious techniques to ensure American intruder training against rail targets mirrored the experiences of the RAF and RCAF over Europe. The training supervisors acquired a supply of phosphorescent sea dye—the material spread in the ocean around pilots or sailors awaiting rescue. The dye is designed to be highly visible in order to facilitate the recovery of personnel in day or night conditions. The pilots undergoing intruder training would intercept trains in southern Florida and conduct mock attacks with their improvised ordnance. The night fighter trainees dropped their loads of sea dye on the trains as they conducted their passes against the engines. The sea dye allowed the pilots to determine the accuracy of their strikes against the trains. Lukas noted that he heard many railroad engineers arrived in Miami after a night run, puzzled about the presence of phosphorescent dye on their engines and box cars.[5]

As the aircrews became more experienced in handling their P-70s and flying in the dark they became bolder, and looked to add some excitement into their training routine. In the spring of 1943, future 419th NFS pilots Second Lieutenant Donald "Don" Dessert and Second Lieutenant Robert "Bob" Winslow were flying two P-70s as a team on a Friday night training mission. Dessert attended college and worked evenings for the Douglas Aircraft Company in the plant that manufactured P-70s before joining the military. He enlisted as an aviation cadet in March 1942, departed for training that September, and completed pilot training in P-38 Lightnings in April 1943. Six future 419th pilots, including Richard "Dick" Stewart, Marvin "Marvelous

Second Lieutenant Don Dessert buzzed a high school football stadium in the Orlando area during a Friday night game, pulling up sharply in order to clear the bleachers and goal posts. Dessert remembered seeing the players scrambling like a herd of cats across the field as the night fighter roared across the stadium. *Courtesy of Don Dessert.*

Second Lieutenant Dick Stewart enjoyed intercepting trains during night fighter training. Stewart would fly to a point a couple of miles in front of the train, turn, and then approach the locomotive at an extremely low altitude. As the pilot roared just above the track he would turn on his landing lights, bringing the terrified train crew to a screeching halt. *Courtesy of Don Dessert.*

Marv" Walker, Frank "Eager Frankie" Hartwick, Don Gardner, Fran Parker, and Dudley Bray completed P-38 training with Dessert. While flying P-61s with the 419th NFS in the Pacific Theater he would later bag one confirmed Japanese Dinah night fighter in October 1944. Winslow, a slender young man with a thick head of hair, would later be credited with a "probable kill" of a Betty bomber while flying from the island of Morotai, off the northern coast of New Guinea, in December 1944.

Dessert and Winslow observed the bright lights of an Orlando area high school football stadium during their evening patrol. Looking for some excitement, the pair banked their aircraft toward the game below them. The two officers dove their P-70s toward the well lighted football stadium, pulling up in order to clear the bleachers and goal posts. Dessert remembers seeing the players scrambling like a herd of cats across the field as the night fighters roared across the stadium. Reports of the incident reached Orlando Air Base, where the commander demanded to know which pilots were responsible for disrupting the football game. No one claimed responsibility or knowledge of the event, and the commander then grounded the entire training squadron.[6]

Second Lieutenant Dick Stewart completed P-38 training with Dessert before transferring with him to the night fighter program in Florida. Stewart, from Canton, Ohio, was the youngest of the 419th NFS pilots and looked like a young kid compared to many of the other pilots. Despite his youthful appearance, Stewart emerged as one of the top pilots of the 419th NFS, bagging two kills (one confirmed and the other unconfirmed) against Japanese Mitsubishi G4M Betty bombers. He also stands out as the only original 419th NFS pilot to have downed a Japanese aircraft while flying a P-38 fighter on a night intercept mission.

Stewart seemed to enjoy intruder training. He related to the author years after the war that he liked to intercept trains passing through the Everglades and other clear areas of Florida. After determining the direction of the train, Stewart would fly to a point a couple of miles in front of the train, turn, and then approach the locomotive at an extremely low altitude. As the pilot roared just above the track he would turn on his landing lights. He often chuckled remembering the incidents, wondering what may have raced through the minds of the rail engineers as they watched the unidentified lights heading toward their engine. Stewart would pull up and over the train as the conductors brought the locomotive to a screeching halt...sparks shooting all along the track from the sliding rail cars.[7]

Night fighter trainees often demonstrated they were a unique breed of pilots capable of many forms of pranks and dangerous aerial feats. Yet, it took this same unique breed of pilot to perform the tasks demanded of them by the USAAF. Not every pilot was capable or willing to engage the enemy in the night sky where the opportunities for fatal mistakes were greatly multiplied. Landing an aircraft during the day could seem routine; landing at night often resulted in accidents. Many pilots died during training accidents in Florida and many more met untimely deaths overseas. While all pilots in low flying aircraft faced the hazards of buildings, trees, or hills along their path, night fighters added the hazard of darkness. It was one thing to dodge the changes in terrain or man-made structures below you; it was quite another to do it at night. Sergeant Harold Young, an aircraft mechanic

P-70 engine maintenance in Florida. The P-70 carried two R-2600-11 engines, which left it underpowered for its interceptor mission. Mechanic Sergeant Harold Young of the 419th NFS often removed pine needles from the engines following low altitude night training missions in Florida. *Courtesy of Leslee Lukas Mays.*

with the 419th NFS, recalled pulling pine needles from the P-70 engines after training missions.[8]

Dessert nearly lost his life while undergoing night fighter training in Florida. Intruder training included missions where the pilots would fly out to sea, navigate across the open water to a coastal town, and simulate bombing a target. Pensacola and Fort Lauderdale, along with their unsuspecting inhabitants, served as the targets of many P-70 night intruder raids during 1943. Dessert simulated an anti-shipping attack in the Fort Lauderdale harbor on one mission. Flight Officer Lane Thompson flew with Dessert on this training mission as the R/O. The pair roared into the Fort Lauderdale area at a low altitude to initiate the simulated attack pattern. As Dessert pulled up on the aircraft he struck an aerial power cable. The impact removed two feet from his left wing and resulted in an electrical blackout for one area of Fort Lauderdale. The two returned to Kissimmee and checked the P-70 for damage. Dessert believed the power cable had struck his right wing, and when he did not see any damage at that location he felt relieved and left the area without reporting the incident. He did not think to look at the other wing, where two feet of its surface was missing. Fortunately, all further night missions were cancelled for the evening, ensuring that another pilot would not face possible problems from the damaged wing. Although informed he would face a court martial for not reporting the damage to his aircraft that night, the charges against Dessert were dropped somewhere along the chain of command.[9] Dessert and Thompson became a pilot-R/O team in Florida and flew many combat missions together in the Pacific Theater. Dessert remembered Thompson as a well educated man and often called him "Rudy Vallee," because he had the tendency to sound like the singer as he belted out tunes while sitting in the rear R/O compartment of the P-61s overseas.

Six of the original 15 pilots of the 419th NFS served with the RAF or RCAF before transferring to the USAAF in 1942. These men brought combat night fighting experience to the USAAF, and their presence boosted the morale of the newly trained pilots joining them in the 419th NFS. Psychologically, it is reassuring to go into combat for the first time with someone who has already proven himself in that arena. Their personal stories offer interesting preludes to the type of pilots they would become with the 419th NFS overseas.

First Lieutenant Al Lukas transferred into the 419th NFS only weeks prior to the unit's deployment overseas. Flight Officer Sanford "Sandy" Bauman, an R/O with the 419th NFS, remembered First Lieutenant Lukas as a "true leader"; someone who held the unit together during a crisis.[10] Lukas often referred to himself as a "vagabond from Chicago" who dreamed of seeing other countries and living the adventures in the books he frequently brought home from the local library. Serving as a private in the U.S. Marine Corps Reserve and attending Northwestern University did not satisfy his yearning for adventure, and books did not substitute for actually traveling overseas. In 1940, Lukas decided to go Absent Without Leave (AWOL) from the U.S. Marine Corps Reserve and traveled to New Orleans with two friends to join the Allied struggle against Germany. Neutrality laws discouraged Americans from traveling into war zones, so the three young men approached the British consul for assistance. One of the friends dropped out of the escapade, but Lukas and Gene Pendleton, the other Chicagoan, assumed non-American sounding names (brothers Harry and Larry Hamilton from Toronto) and received phony papers listing them as British seamen. The pair sailed on the Norwegian oil tanker *John Knudson* to Britain and then back to Aruba for another shipment of oil bound for France.

In Le Harve, the duo departed the ship and traveled to Paris, where they joined a newly formed "*Lafayette Escadrille*" type unit of American pilot trainees. Like its famous World War I counterpart, the French unit consisted of American volunteer pilots. However, France fell to the Germans before the pilot candidates could make their first solos. Lukas and his buddy received hasty discharges from the French Air Force, stole bicycles, and made their way to Marseilles, where they hopped a ship sailing for the navy port of Toulon. The two Americans then stowed away on another ship, this time bound for Oran, Algeria, only to arrive in time for the British attack on the Vichy French fleet in the harbor. This was the same raid where Flight Lieutenant Kuttelwascher refused Vichy French orders to attack the British task force. After escaping to shore, the two engaged in a series of adventures across North Africa. Posing as Arabs, they were arrested at the Moroccan border as suspected smugglers. Released by an anti-German police officer, the two men continued their journey—now masquerading as discharged French Foreign Legionnaires. Years later when asked how he acquired French Foreign Legion discharge papers, Lukas grinned and noted that he and Gene met two Legionnaires who could not hold their liquor!

In Morocco, the duo were befriended by "ladies of the evening" and hidden in the Casbah—an area the police tended to avoid. The pair later stowed away on a trawler carrying olive oil bound for France, knowing that the British would halt and inspect the contents of the vessel before allowing it to continue. A British navy inspection

First Lieutenant Al Lukas served five Allied countries during World War II. He signed on with the Norwegian merchant marine and made three voyages between North/South America and Europe on the oil tanker *John Knudson* in 1939 - 1940. He then briefly joined the French *Armee de l'air* before having to escape to Algeria following the fall of France. After a series of adventures across North Africa and stowing away on a freighter, he emerged during a Royal Navy inspection of the vessel to falsely declare himself a Canadian survivor of a torpedoed ship. Lukas joined the Royal Air Force (RAF) and flew in a Royal Canadian Air Force squadron before transferring to the United States Army Air Force at the end of 1942. Note the RAF wings on his uniform. *Courtesy of Leslee Lukas Mays.*

First Lieutenant Howard Daniel (far left), a former pilot in the Royal Air Force and Royal Canadian Air Force, joins other members of the 419th NFS in sharing a tub of "jungle juice" or "purple passion." First Lieutenant Don Dessert sits immediately to the right of Daniel in the photo. Squadron members made the alcoholic drink from whatever source of alcohol available—often "cheap gin" brought back from air crew leave in Australia or New Zealand. Any available fruit juice in the local post exchange would be added to the liquor. With grape juice being frequently in stock, the concoction had a purple color, earning "purple passion" as one of its nicknames. First Lieutenant Bob O'Brien remembered that it "tasted like bootleg whiskey, but smoother." Dessert noted they always had a sock in the bottom of the container—a tradition maintained at many Army functions today. *Courtesy of Don Dessert.*

near Gibraltar provided the opportunity for the pair to surface and claim that they were Canadian seamen who had survived a torpedo attack on their ship. The British bought their story and gave them passage to England. Lukas and his companion joined the RAF after arriving in London. Due to his exceptional night vision, the RAF offered Lukas the opportunity to join a night fighter unit numbered as an RCAF squadron. The Chicagoan agreed, while his friend accepted an offer as a day fighter trainee. Lukas joined Number 418 Squadron of the RCAF and went on to score the unit's first night fighter kill. The Germans shot down Gene on his first mission over the continent, and he remained a prisoner for the duration of the war.

Lukas returned to the United States in March 1943 after three years in Europe in order to help train America's new night fighter pilots. Stateside duty as a training officer did not suit the active lifestyle of Lukas, who quickly became bored with pushing paper and training new pilots in night fighter tactics. After an outing of buzzing topless co-eds sunning on dorm roofs

The 419th NFS' first Commanding Officer, Captain John McCloskey of Philadelphia, flew de Havilland Mosquitoes with the Royal Air Force and was credited with one confirmed and one probable night fighter kills in duels with *Luftwaffe* bombers over Great Britain. McCloskey died in a tragic training accident within weeks of arriving at Guadalcanal. *Courtesy of Leslee Lukas Mays.*

First Lieutenant John Laffey served as a Royal Air Force (RAF) test pilot before transferring to the United States Army Air Force (USAAF) night fighter program. RAF and USAAF flight logs found after his death in the Philippines revealed that the Chicagoan had flown over 50 different variants of British and American aircraft during his military service. *Courtesy of Leslee Lukas Mays.*

and landing a small plane on a golf course for a drink at the club bar, Lukas and First Lieutenant "Pappy" Harper celebrated 4 July confined to the barracks as the staff duty officers. After forgetting to turn off the light in his quarters, Lukas took his pistol and simply shot it out. The USAAF offered Lukas the choice between a combat tour and a court martial. Lukas wrote in his flight log "I decided to go overseas again. Can't stand the war in Florida." Lukas joined the 419th NFS three weeks before the unit deployed to the Pacific Theater to serve with the 13th Air Force.[11]

The 419th NFS Commanding Officer, Captain John McCloskey of Philadelphia, flew de Havilland Mosquitoes with the RAF and was credited with one confirmed and one probable night fighter kills in duels with *Luftwaffe* bombers over Great Britain. McCloskey exerted an almost charismatic influence over the 419th NFS personnel, who often tapped into his inner strength and previous military experience as they prepared to enter the war zone in the Solomon Islands area. Sergeant Harold Young remembered McCloskey as a man to whom he looked up with considerable pride. The Captain was a leader who inspired others to be their best.[12] McCloskey's death, the first casualty of the 419th NFS in the Pacific Theater, stung the entire unit.

First Lieutenant Howard Daniel hailed from Blue Mound, Kansas, served as the captain of his high school basketball team, and worked as a part time ambulance driver while attending Northwest Missouri State Technical College. Described by friends as a "happy go lucky" yet modest person, Daniel joined the RCAF in the fall of 1940 and attended pilot training in Ontario. A crash landing during training resulted in his hospitalization for six months. However, Daniel recovered and volunteered for duty in Great Britain. Transferred to the RAF, Daniel flew Supermarine Spitfires for a night fighter squadron. His most memorable mission occurred in early 1942 when damage to his aircraft forced him to make an emergency landing on a heavily mined strip of beach.

Daniel successfully landed his damaged Spitfire on the beach and walked away without detonating any mines. British military authorities were so impressed that he survived the ordeal they awarded him a commendation "for extreme bravery in the course of duty." Daniel as well as Lukas flew air missions in support of the Canadians during the 1942 Dieppe Raid. First Lieutenant Robert "Bob" O'Brien, the 419th NFS Ordnance officer, fondly remembered Daniel as the pilot who would often take him up for hops in the squadron's A-24, the Army's variant of the Navy SBD dive bomber. The 419th aircrews utilized two Douglas A-24 Banshees for radar observer training and hops between airfields. O'Brien reminisced, "Daniel would take the A-24 up to altitude, tell me the compass heading, and have me take the controls. He then removed his shirt to sunbathe in the front seat."[13] Daniel failed to return from a dawn P-61 patrol off New Guinea in October 1944.

First Lieutenant John Laffey of Chicago served as an RAF test pilot before transferring to the USAAF night fighter program. Flight Officer Bauman remembered Laffey as a "heavy gregarious guy," loved by all in the unit.[14] As a former test pilot, Laffey could not resist getting into any type of aircraft that he had never flown. This curiosity would contribute to his untimely death in 1945. 419th NFS veterans assert that a review of Laffey's RAF and USAAF flight logs after his death revealed that the Chicagoan had flown over

First Lieutenant Charles Hicks (standing in the center of this trio with his Radar Observer Flight Officer John Radzieta on the left and Gunner Sergeant William Yahn on the right) dropped out of Penn State and crossed the border into Canada, where he joined the Royal Canadian Air Force (RCAF) early in World War II. After completing pilot training, Hicks volunteered for service in Great Britain and became an instrument pilot instructor. Like the others, Hicks transferred to the United States Army Air Force in 1942. Hicks was the only former RAF/RCAF pilot among the original 419th NFS aircrews to complete the American night fighter school in Florida before being assigned to the squadron. *Courtesy of Don Dessert.*

Former Royal Air Force veteran First Lieutenant William Mackie of Tampa Bay, Florida, sitting in a Guadalcanal foxhole. First Lieutenant Bob O'Brien seized an opportunity to fly with Mackie while the unit was undergoing P-61 Black Widow transition training on Guadalcanal. "What a hot dogger. We flew out of the bay then came back and he took the P-61 down on the deck – I mean we flew past a freighter in the harbor and we were so low that as we passed the stern, I had to look up to see the sailors on deck." *Courtesy of Leslee Lukas Mays.*

50 different variants of British and American aircraft during his military service.

First Lieutenant Charles Hicks dropped out of Penn State and crossed the border into Canada, where he joined the RCAF early in World War II. After completing pilot training, Hicks volunteered for service in Great Britain and became an instrument pilot instructor. Like the others, Hicks transferred to the USAAF in 1942. Hicks was the only former RAF/RCAF pilot among the original 419th NFS aircrews to complete the American night fighter school in Florida before being assigned to the squadron. Flight Officer John Radzieta flew as Hicks' R/O throughout their overseas service:

> "I am proud to say that Hicks handpicked me to be his R/O. He was a gentleman, and my wife felt good that I had such a fine man for a pilot. When we got to the 'Canal' he was the eager one. He did not wait for things to happen; he made them happen."[15]

The sixth former RAF/RCAF veteran was First Lieutenant William Mackie of Tampa Bay, Florida. Flight Officer Bauman remembered Mackie as a "walking encyclopedia"—a man who constantly read anything within reach and seemed to never forget anything. First Lieutenant O'Brien seized an opportunity to fly with Mackie while the unit was undergoing P-61 Black Widow transition training on Guadalcanal:

"What a hot dogger. We flew out of the bay then came back and he took the P-61 down on the deck—I mean we flew past a freighter in the harbor and we were so low that as we passed the stern, I had to look up to see the sailors on deck. Mackie forgot about the palm trees lining the shoreline and had to tip the plane on its wing to pass through them. I was too young to ride with him again."

Mackie would return home following what every 419th NFS veteran describes as an unjust court martial on Guadalcanal.[16]

These six ex-RAF/RCAF officers assumed the initial command positions of the squadron when it deployed overseas. McCloskey served as the squadron's first commanding officer and Daniel became the squadron Operations Officer, with Lukas as his assistant. Laffey, Mackie, and Hicks were the first squadron flight leaders. Lukas would later claim the first kill of the squadron (the second in the entire Pacific Theater with a P-61) when he downed a Japanese Sally bomber near Noemfoor in August 1944. In a sad testament to the high attrition suffered by American night fighter pilots in World War II, half of these men (McCloskey, Daniel, and Laffey) would not make it home again to the United States after deploying with the 419th NFS in late 1943. On October 26, 1943, the 285 officers and enlisted men of the 419th NFS sailed from San Francisco, California, to replace Detachment B 6th NFS as the primary 13th Air Force night fighter asset. They were destined for a dangerous adventure that would take them through the Solomon Islands, New Guinea, and the Philippines in support of the 13th Air Force. Some would never return.

419th NFS personnel enjoy a final meal in San Francisco before departing for Guadalcanal. On the far left, from front to back, are First Lieutenant Charles Hicks, Second Lieutenant Bob Winslow, and Flight Officer John Radzieta. On the right from front to back are First Lieutenant John Laffey, Flight Officer Ed Kelly, Second Lieutenant Dick Stewart, and Flight Officer Anthony Grotzinger. *Courtesy of John Radzieta.*

Second Lieutenant Don Gardner with a Douglas P-70 Nighthawk. Gardner became the first 419th NFS pilot to be shot down in combat. The P-70, a variant of the A-20, like the Royal Air Force's Boston bomber, was America's first night fighter. The Army Air Force planned for the P-70 to serve as an interim night fighter until the fielding of the Northrop P-61 Black Widow. *Courtesy of Leslee Lukas Mays.*

2

The Solomon Islands (1943)

"[The Japanese soldiers] twisted and turned as the bullets passed through them. I was so intent on firing at the patrol that I almost forgot about pulling up the P-38 before hitting the ground. I yanked on the stick and managed to clear the area."

First Lieutenant Richard (Dick) Stewart
pilot, 419th NFS

Allied forces were on the defensive during the first half of 1942 in the Pacific Theater. Japanese forces overran the Philippine Islands, Dutch East Indies (modern Indonesia), French Indochina (modern Vietnam, Cambodia, and Laos), Malaysia, Singapore, and other areas. Their troops were moving westward across Burma towards India and consolidating positions in China. In the Southwest Pacific, Japanese forces engaged Allied troops on New Guinea and were seizing bases in the Solomon Islands with the aim of cutting the lines of communication between the United States and Australia. The Solomon Islands, located east of New Guinea and stretching toward the southeast and Australia's eastern coast, would see the first major American counter offensive of the Pacific Theater.[1]

The Japanese military planned to occupy a string of bases in the Solomon Islands to support their operations in eastern New Guinea and movement toward the Coral Sea and islands of the southeast Pacific. On the island of New Britain, the Japanese established a major naval and air base at the harbor town of Rabaul. Moving southeast from New Britain, bases were placed on Bougainville Island. A channel of the sea ran from Bougainville Island in the north to Guadalcanal in the south. Each side of this channel, known as the Slot, was lined with islands of different sizes. The Japanese built airfields on some of these islands as they moved southward from Bougainville toward Guadalcanal. Other islands became homes for float planes based in bays. Meanwhile, the halting of a Japanese aircraft carrier-led task force at the Battle of the Coral Sea in May 1942 helped signal that the United States was prepared to turn the tide in the war with Japan.

When the Japanese began construction of an airfield on Guadalcanal in July 1942, the United States decided that this would be the location where it would draw a "line in the sand." American forces landed on Guadalcanal and the nearby islands of Florida and Tulagi on 7 and 8 August. American Marine and Army troops fought a tough campaign with limited resources, including air power, as they struggled to first remain on Guadalcanal and then drive the Japanese defenders from the island. By the end of 1942, it was finally evident that the United States was on Guadalcanal to stay.[2]

1943 emerged as a year of attrition as the United States and its allies, located at the bottom of the Slot,

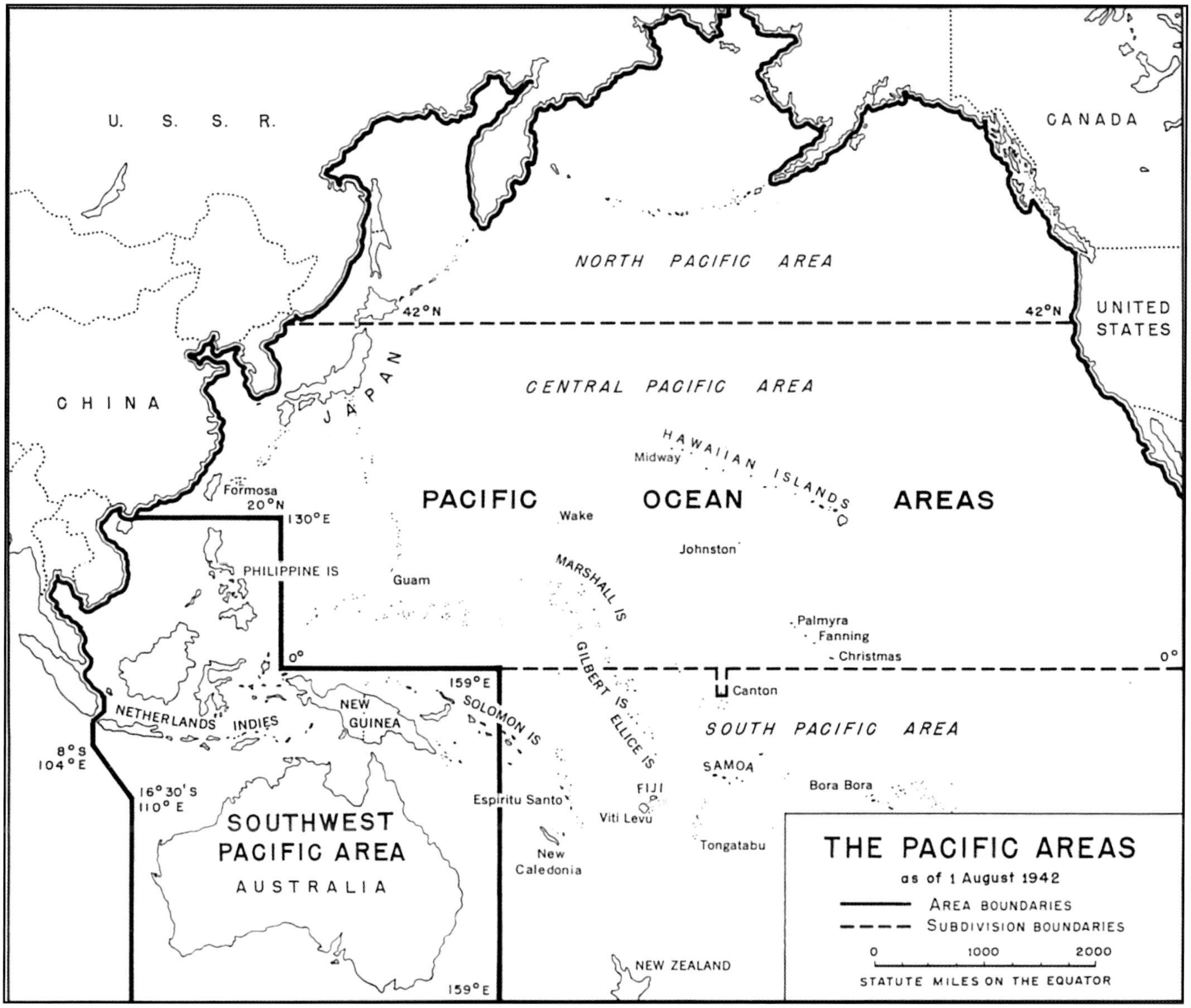

Pacific Command boundaries. *Courtesy of the Center for Military History.*

faced the Japanese who held the rest of the channel and its islands leading to Bougainville. To reach the major base area of Rabaul, Allied forces needed to battle their way up the Slot and seize Bougainville; a task that would take 11 months. 1943 also witnessed the birth and growing pains of the 13th Air Force.

The American Joint Chiefs of Staff divided the Pacific Theater into two combat command areas. Army General Douglas MacArthur led the Allied Forces in the Southwest Pacific Area, which included Australia, New Guinea, most of the Dutch East Indies, and the Philippine Islands. American Army Air Force elements in the Southwest Pacific were organized under the 5th Air Force and commanded by Lieutenant General George Kenney. The rest of the Pacific Theater became the responsibility of Admiral Chester Nimitz. Plans called for a two-pronged offensive against the Japanese, with MacArthur stabilizing the situation in the Southwest Pacific and initiating a rollback of the Japanese to the Philippines, while Nimitz led an amphibious based assault across the islands of the Central Pacific. A combat sub-command known as the South Pacific Area was established under Nimitz to work in cooperation with General MacArthur. Guadalcanal and Santa Isabel Island above it marked the western boundary of the South Pacific Area. To the west lay the Southwest Pacific Area under the leadership of General MacArthur. As MacArthur's forces battled the Japanese on New Guinea, the Allied forces of the South Pacific Area initiated an offensive through the Solomon Islands following the seizure of Guadalcanal.

Japanese Mitsubishi F1M Pete. The Pete and other small fighters flying night harassment missions over Guadalcanal were nicknamed "Louie the Louse" by American troops, who also nicknamed any of the Japanese large bombers "Washing Machine Charlie" due to the sounds of their unsynchronized engines. *Courtesy of Bob Fowble.*

Although Vice Admiral Robert Ghormley assumed initial command of the South Pacific Area, Admiral William "Bull" Halsey held the post when American forces landed on Guadalcanal in August 1942.[3]

As commanders in the other combat areas, Halsey led a mixture of Army, Army Air Force, Navy, Marine Corps, and Allied forces. Although a subordinate of Nimitz, Halsey also operated as a tactical subordinate to MacArthur. In a 1947 *Saturday Evening Post* article, Halsey explained his tender situation between two commanders:

"This arrangement was sensible and satisfactory, but it had the curious effect of giving me two 'hats' in the same echelon; my original hat was…under Nimitz, who controlled my troops, ships, and supplies; as soon as I crossed the 159th meridian, I gained another hat, under MacArthur, who would now control my strategy."[4]

Army Air Force units in the South Pacific lacked a central Army organization to control and coordinate their efforts. In February 1943, the 13th Air Force was formed to provide a command structure for the various Army Air Force elements in the South Pacific Area. These air units included the Army Air Force elements of the famous "Cactus Air Force," which had defended Guadalcanal since the arrival of the first American planes at Henderson Field. Major General Nathan Twining assumed command of the newly established 13th Air Force and provided the direction it needed to defend Guadalcanal and other islands in the southern Solomons from Japanese air and naval forces. Twining established his headquarters at Espiritu Santo, in the New Hebrides, a chain of islands located southeast of the Solomon Islands. The Allies established another organization known as the Commander, Air, Solomon Islands (COMAIRSOLS) to provide coordination between the 13th Air Force and the air elements of the Navy, Marine Corps, and New Zealand as the forces of the South Pacific Area prepared to advance northward through the Solomon Islands in cooperation with the Southwest Pacific Area offensive on New Guinea.[5]

1943 was also the year that American night fighters entered the war against Japan with the 13th Air Force. From the first days of the American landing in August 1942, Japanese aircraft raided Henderson Field and other American held areas on Guadalcanal during the night. Small float planes from Rekata Bay, on Santa

Isabel Island, and medium bombers from Bougainville made regular appearances over Guadalcanal. The planes rarely came out in large numbers and normally attacked as single aircraft or pairs. They also tended to drop single bombs or partial bomb loads rather than the entire cache in their bomb bays or hanging on their wings during a single pass. Instead, the Japanese air crews purposely divided their bomb loads and dropped the ordnance over several passes spaced 10-15 minutes apart. The bombs, often dropped from high altitudes, did not cause much damage, except for occasional chance hits that did result in destroyed and damaged aircraft. Sometimes an ammunition or fuel dump would erupt in flames when a bomb fell within its perimeter. The impact on morale far outweighed the materiel damage caused by the night raiders. In particular, pilots suffered from the lack of sleep after flying long tiring combat missions during the day, and the attacks zapped the morale of everyone on the ground. By attacking as single aircraft or pairs every 10-15 minutes during a raid, the Japanese ensured they were over Guadalcanal for long stretches of time, preventing the Americans on the ground from getting the rest they needed to be fully effective in combat during the day. In addition, the time spent in fox holes during the night raids increased the exposure of the men to malaria carrying mosquitoes.[6]

Two-engine Japanese bombers (Mitsubishi Ki-21 Sally and Mitsubishi G4M3 Betty) were collectively referred to as "Washing Machine Charlie" because American soldiers compared the sound of the unsynchronized Japanese aircraft engines droning overhead to the sounds made by old mechanical washing machines back home. Thus, each large raider was known as "Washing Machine Charlie" regardless of what type of bomber passed overhead. Small single engine float planes (primarily the Mitsubishi F1M1 Pete) were collectively known as "Louie the Louse" when they raided Guadalcanal or other American-held areas in the Solomon Islands.

Not all of the ordnance dropped by "Washing Machine Charlie" or "Louie the Louse" during the raids involved explosives. Many times, the Japanese dropped empty glass bottles from their aircraft. Reports passed between Americans in the era claimed that the wind whistled into the bottles as they fell and could be heard by those on the ground just prior to striking the earth, where they shattered. Supposedly, an empty bottle had just as much morale damaging impact as a bomb due to its sound in flight and when shattering on the ground. Many American crews, including those of Detachment B 6th Night Fighter Squadron (NFS) and the 419th NFS after the latter's arrival on Guadalcanal in November 1943, returned the favor by dumping empty beer bottles from their aircraft over Japanese targets at night whether or not they could be heard on the ground. American P-70 Nighthawks carried small bomb loads, preventing their crews from striking multiple targets during night intruder missions. Their former bomb bays were redesigned at belly tubs for guns. Second Lieutenant Jim Allen of the 419th NFS recalled an innovative solution to the problem of small bomb loads:

"The P-70s could not carry much ordnance, and by the time we hit Japanese areas up the Slot we would be out of bombs and could not hit their bases located along the route back. We had received a nice ration of Australian beer, and their bottles were twice as large as ours. One of the guys said that he heard a bottle coming from a fairly high altitude sounded like a bomb coming down. We loaded two or three cases of beer bottles—empty, of course—and on return trips would throw them out over Japanese camps. We never did know if this worked, but it sure was fun to throw them out."[7]

Army Major General Alexander Patch (who had assumed command on Guadalcanal from Marine Corps Major General Alexander Vandegrift in December 1942) informed Admiral Halsey upon taking command that he needed six designated night fighters to eliminate the Japanese raiders flying over Guadalcanal. The request was relayed to officials in Washington, D.C., noting that the area required a night fighter squadron with Ground Control Interception (GCI) radar. Night fighter aircraft would not be able to reach Guadalcanal until sometime in February 1943—two months after Patch's request.

During the waiting period, many day fighter pilots itched to make a night interception in retaliation for the nightly raids over Guadalcanal. A Marine Corps Grumman F4F Wildcat attempted to intercept a Japanese night intruder one evening over Guadalcanal but was forced to break contact when American anti-aircraft artillery filled the sky with 90mm shells. Another night an Army Air Force P-38 went up to bag a Japanese bomber. However, the Japanese plane simply remained clear of any searchlights and waited for the P-38 to land. The Japanese pilot then returned and bombed the airfield. After considerable pleading, the 339th Fighter Squadron received permission to formally attempt intercepts of the enemy night intruders. On January 29, 1943, Captain John Mitchell launched just prior to dawn and pursued a Japanese twin-engine bomber over

Guadalcanal. Mitchell caught the enemy aircraft and flamed it. The bomber crashed into the sea to the sheer delight of everyone on the ground who observed the engagement. (Mitchell is best known as the officer who led the mission that surprised and shot down the Betty bomber carrying Japanese Admiral Isoroku Yamamoto on April 18, 1943.) Despite the glee over the downing of "Washing Machine Charlie," the kill was a rare exception and was counter-balanced by many failed attempts of the day fighter pilots to repeat the feat.[8]

In response to the request for night fighters to be assigned to the 13th Air Force, six P-70 Nighthawks (the unit was designated simply as "Night Fighter Squadron" and later the "6th Night Fighter Squadron") arrived on Guadalcanal in February 1943. At this time in the war there was only one night fighter squadron in the entire Pacific Theater. This was the 6th Fighter Squadron, which began training as a night fighter unit after the bombing of Pearl Harbor in December 1941. The unit flew with Curtis P-40 Warhawk aircraft until receiving the newly converted P-70 Nighthawks.

Six P-70s departed Hawaii on February 18, 1943, escorted by LB-30s, the transport version of the B-24 Liberator bomber, which provided the long range navigation and carried the ground crews. One P-70 disappeared in a rain squall encountered along the route. The night fighters, officially designated as Detachment B, Night Fighter Squadron, established operations at Carney Field, Guadalcanal, with five P-70s amid considerable anticipation of everyone who desired to see the end of "Washing Machine Charlie" and "Louie the Louse." At the same time, P-70s flew from Hawaii to New Guinea to provide night fighter defense for the 5th Air Force. This unit would be referred to as Detachment A.

Employment of the P-70s and their tactics developed from a combination of British lessons in the European Theater and simple "trial and error." Basic tactical employment scenarios emerged after the arrival of the P-70s in early 1943. The first involved posting a P-70 approximately 50 miles from Guadalcanal along a route known to be traveled by Japanese aircraft

Douglas P-70 Nighthawks of Detachment B, 6th Night Fighter Squadron (NFS), buzz Guadalcanal. Detachment B made one confirmed kill with a P-70 flying from Guadalcanal. The P-70 proved to be underpowered and could not reach the altitude of most Japanese bombers. Detachment B carried two names while on Guadalcanal. Until September 16, 1943, it was known as Detachment B, 6th NFS. After this date its name changed to Detachment B, Night Fighter Squadron, while it awaited the arrival of the 419th NFS, with which it would merge. Due to the name change late in its 13th Air Force mission, this book refers to the unit as Detachment B, 6th NFS to prevent any confusion with the various times the 419th NFS formed a Detachment B. *Courtesy of Leslee Lukas Mays.*

419th NFS P-38 Lightning. The 419th NFS inherited its P-38s from Detachment B, 6th NFS, which made eight kills of Japanese bombers and fighters with these aircraft. *Courtesy of Dot McConnell.*

raiding the islands. The P-70 would orbit the area in the dark, awaiting a radio call from the GCI controller on the ground. The controller would report a "bogey," or unknown aircraft, in the area and provide the P-70 pilot with the altitude, distance, and vector (compass heading) of the bogey displayed on the GCI radar. The bogey could be a Japanese raider or a friendly aircraft with a malfunctioning IFF (Identification Friend or Foe) box. The latter was a classified device in each Allied aircraft that emitted a signal indicating to radar sites and others with receivers that the plane was friendly and not enemy. The P-70 pilot followed the directions provided by the GCI controller until his on board Radar Observer (R/O) made contact with the short range radar set in the Nighthawk. At this point the R/O assumed responsibility for directing the pilot. Ideally, the R/O sought to bring the P-70 behind the bogey, where the American night fighter would be harder to spot at night. The pilot carefully followed the directions of his R/O and scanned the night sky based on his directions. When locating the bogey, the pilot had to visually identify the plane as either friend or foe. An attack could not be made without a positive identification that the bogey was a Japanese aircraft. The P-70 was cleared to hunt the enemy aircraft until the latter flew into the anti-aircraft circle. This was an imaginary boundary, normally extending 10 miles from Henderson Field, where anti-aircraft guns would assume the primary responsibility for engaging the enemy raider. The P-70 continued to orbit outside of the anti-aircraft circle, waiting to see if the Japanese plane emerged safely from the ring of guns. If so, the P-70 renewed the chase.

Later the P-70s hunted in cooperation with P-38 fighters. If a bogey managed to elude the P-70 and enter the anti-aircraft circle, a P-38 fighter orbiting the area had the option of calling a halt to the anti-aircraft artillery and engaging the raider if it had been spotted by ground based searchlights. In 1944, First Lieutenant Dick Stewart of the 419th NFS bagged a Japanese Mitsubishi G4M3 Betty bomber in this manner, although the anti-aircraft guns on Middleburg Island did not halt their barrage when requested. Another employment scenario involved having the P-70s orbit near Henderson Field awaiting the illumination of a Japanese raider by searchlights or the acquisition of a bogey on the Nighthawk's airborne radar system. This tactic emerged after the realization that the P-70 was not an effective night fighter. However, it is the tactic by which Detachment B scored its only confirmed night kill with the P-70.

The P-70 Nighthawk did not live up to its name when employed in combat and suffered many problems. General Twining himself listed the limitations of the P-70s during combat. First, the ceiling, or maximum altitude, of the P-70 was 22,000 feet. Japanese bombers normally attacked at altitudes above 22,000 feet, preventing the P-70 from being able to reach them. Second, the P-70 had an inadequate rate of climb. It took approximately 45 minutes for the Nighthawk to reach its maximum ceiling of 22,000 feet. A bogey could easily depart the area if pursued by a P-70 struggling to reach a higher altitude. Third, the maximum speed of the P-70 at its maximum altitude was often too slow to catch its prey. In the early hours of May 16, 1943, a Japanese Betty bomber outran a pursuing P-70 by using a shallow dive to increase its speed. Fourth, the electrical system of the aircraft frequently failed due to overloading. Fifth, the radar system onboard the P-70 tended to malfunction at high altitudes. Ground scatter from the radar waves made the airborne sets nearly useless below 5,000 feet, and low air pressure frequently affected the radar above 8,000 feet.[9]

The first radar assisted American kill of an enemy aircraft by a P-70 occurred over Guadalcanal on April 19, 1943. Captain Earl Bennett (pilot) and Technical Sergeant Raymond Mooney (R/O) were airborne as a Japanese Mitsubishi G4M Betty bomber approached the island.[10] Japanese intruders flew over Guadalcanal throughout the night of 19 April. Bennett made a fruitless two hour sortie in search of the raiders before landing in frustration. He took off again from Carney Field on Guadalcanal at 3:21 AM for a fresh attempt at ambushing an enemy aircraft. A second night fighter manned by Captain Milton Hillebrand (pilot) and Sergeant Dick Ehlert (R/O) also prowled in the dark over Guadalcanal. Bennett and Mooney circled near Savo Island while Hillebrand and Ehlert orbited near Tulagi Island as they waited to either locate a bogey on their radar sets or trapped in a searchlight beam from the ground defenses.

Japanese aircraft periodically flew over Guadalcanal and released ordnance on the American airfields that night. At the time, no one was sure whether there was one Japanese intruder who kept returning over the airfields or multiple enemy planes passing over Guadalcanal in shifts to harass the Americans on the ground. Hillebrand and Ehlert spotted a Japanese aircraft trapped within the beam of a searchlight on several occasions over a 20 minute period but could not catch and down the plane. At one point Hillebrand, while orbiting near Tulagi, spotted a Betty locked in a searchlight beam and turned to intercept the enemy bomber. However, the American defenses on Tulagi heard the engines of the P-70 and began firing anti-aircraft guns at the friendly plane. While dodging their own anti-aircraft fire, Hillebrand and Ehlert lost visual contact with the Japanese bomber over Guadalcanal.

At 4:18 AM American searchlight crews managed to lock onto a Japanese Betty bomber directly over Henderson Field. Bennett and Mooney were orbiting at 22,000 feet near Savo Island and immediately launched a pursuit of the Betty trapped in the searchlight beam. As the P-70 approached the location of the Betty, the Japanese plane managed to elude the searchlights and disappear into the night. The Betty bomber continued to elude the American defenses and entered the 90mm anti-aircraft gun circle around the airfields on Guadalcanal. Ignoring the danger of being tagged by their own anti-aircraft artillery, Bennett and Mooney pursued the enemy intruder. Again, the bomber eluded the searchlights. However, this time Mooney managed to get a radar fix on the enemy plane within 30 seconds of its disappearing act and provided a new vector for Bennett to follow. Bennett spotted the exhaust flames from the Betty's engines and knew he finally had "Washing Machine Charlie" in his sights. The pilot fired the Nighthawk's four 20mm cannon and the Japanese bomber exploded in mid-air.

Tracers from the P-70 could be seen by personnel on the ground when Bennett commenced firing on the Japanese Betty bomber captured in the searchlight beam. Ehlert remembered watching Bennett and Mooney down the Betty:

> "Bennett and Mooney got the Betty right over the island. You could see the tracers leave the P-70 and then watch the Betty explode. All the people who saw the action [from the ground] clapped and cheered and hollered. It was quite a sight…The Major [Sidney Wharton, commanding officer of Detachment B] broke out a bottle of 17 year old Scotch and really hung on a dilly for Bennett and Mooney."[11]

Troops on the island provided Detachment B with a hand painted sign thanking them for the "Jap-less sleep" after the kill. Mooney remained with the 6th NFS throughout the war and did not transfer into the 419th NFS with many of his fellow air crew in December 1943. He received a promotion to the status of Flight Officer and later a commission as a Second Lieutenant with the 6th

NFS in the Central Pacific. Mooney is officially credited as the R/O on three kills of Japanese Betty bombers while flying in P-61 Black Widows.

Despite the first kill of a Japanese bomber by a P-70, the presence of the Nighthawk over Guadalcanal did little more than persuade the Japanese bombers to remain at high altitudes when they passed over the island. Other fighter units continued to try their hands at intercepting the Japanese aircraft utilizing P-38s, since Detachment B had been able to bring down only one bomber with the inadequate P-70s. A low-intensity red light was set up approximately one mile southwest of Fighter Strip No. 2 on Guadalcanal. P-38 pilots circled this area awaiting the arrival of Japanese aircraft during the night. If a Japanese bomber entered the area and was discovered by the ground searchlight crews, the local Fighter Control permitted any orbiting P-38 to have priority over the anti-aircraft guns to shoot down the intruder. These P-38s demonstrated some success at bringing down "Washing Machine Charlie." During the evening hours of 13 May four Japanese bombers crisscrossed Guadalcanal, dropping 11 bombs, including one that ignited a munitions dump. First Lieutenant William Smith of the 12th Fighter Squadron orbited the island that night waiting for an opportunity to pounce on a Japanese bomber. As the ground searchlights illuminated a bomber, Smith would attack the aircraft. He succeeded in bringing down one bomber with a confirmed kill and a second as a "probable" kill.[12]

The full moon the next week resulted in an increased number of Japanese raiders. On 19 May several Japanese bombers took advantage of the full moon and struck Guadalcanal, killing 14 American servicemen and wounding an additional 20 men. However, P-38s were waiting for the Japanese bomber crews that night. Seven P-38s rotated coverage throughout the night to ensure a continuous American presence in the air. Just after midnight, Major Louis Kittel of the 70th Fighter Squadron spotted a Japanese bomber illuminated by a searchlight and moved in for a kill. Kittel maintained visibility of the bomber thanks to not only the searchlight, but also the bomber's exhaust flames being visible as they emerged from the engines. Although the bomber eluded the searchlight crews, Kittel continued the pursuit of the Japanese plane by following the exhaust flames and the trails of tracers fired by the bomber's defensive gunners. He quickly dispatched the Japanese Betty bomber and within seconds downed a second Betty. It had been a good night for the defenders of Guadalcanal.[13]

Despite being with the 13th Air Force since February 1943, Detachment B had downed only one enemy aircraft with the P-70. The Nighthawk was simply not the right aircraft, though the Army Air Force awaited the fielding of the highly anticipated and classified P-61 Black Widow, which was expected before the end of 1943. The only temporary solution was to employ more P-38s in the night fighter role. Although the P-38 was not the best solution either, the aircraft had certainly proven more reliable at knocking Japanese bombers from the night skies than the P-70. Detachment B flew its final long-range interception mission with the P-70 on 20 May. After this night, the P-70 would only fly low-level interceptions in cooperation with GCI radar while P-38s would assume the major night defense burden over Guadalcanal. The unit borrowed P-38s from the other squadrons on the island and began re-training its pilots to fly the fighters during night missions.[14]

Detachment A 6th NFS in New Guinea scored the second and last official victory with a P-70 over a Japanese night intruder. Although Detachment A supported the 5th Air Force in New Guinea rather than the 13th Air Force, this particular engagement is an important element of night fighter history, since it represents the only additional successful intercept and kill of an enemy aircraft by a P-70 crew. During the night of May 15, 1943, a Japanese Mitsubishi Ki-21 Sally bomber attempted to slip through the American air defenses in New Guinea. Normally, Japanese aircraft remained at high altitudes to avoid the American night fighters defending Port Moresby as they did around Guadalcanal. Poor weather forced the Sally pilot to maneuver his aircraft to a lower altitude of 18,000 feet. A P-70 piloted by Captain Burnell Adams with Flight Officer Paul DiLabbio serving as the R/O scrambled that evening in response to a reported incoming bogey. DiLabbio reported:

"I picked up the first bogie at approximately 18,000 feet range and quite lower than our altitude of 26,000 feet. I gave Lieutenant Adams directions to dive and turn port. At this time, our intercom started to act up and I was able to give only one more set of instructions which were to level out the dive and steady the heading. Immediately, Adams said he had a visual and seconds later I heard the 20 mm guns being fired. Adams scored several hits and the Jap bomber exploded. We identified it as a 'Sally.'"[15]

As the Sally bomber crashed DiLabbio acquired a second bogey on his radar scope, but was not able to communicate the information to Adams due to a complete failure of the intercom system. The crew firmly believed they could have located and downed the second bomber had their intercom system functioned. Later Detachment A learned that the Japanese aircraft had monitored the radio transmissions of the night fighters that evening and heard the excited chatter about the Sally flying at a low altitude. After that night, the Japanese bombers flying in the vicinity of Port Moresby never dropped below 28,000 feet.[16] Adams later transferred to the 80th Fighter Squadron rather than remain with the night fighter program and join the newly arrived 418th NFS. The 80th Fighter Squadron included Captain Richard Bong, the highest scoring American ace of World War II with 40 kills. Adams would score six more kills while piloting P-38s with this unit.

Although the P-70 is officially credited for only two kills during World War II, there is evidence to support one "probable" and one "possible" kill as well—both with Detachment B in support of the 13th Air Force. A "probable" kill reflects an enemy aircraft that may have been shot down but the proof is lacking. Proof that would verify a "probable" kill and give the crew official credit could include a second witness to the event or evidence of a crash site. Official Detachment B documents credit Captain Emerson Barker with the probable kill of a Japanese Betty bomber on August 13, 1943, in the vicinity of Guadalcanal. However, there are no further official records that provide any details of the encounter.[17] The term "possible" is not an official designation of the Army Air Force, but is utilized here to distinguish an engagement where an enemy aircraft is believed to have been hit and destroyed but the available evidence is not strong enough to grant the crew a "probable" kill.

Japanese forces were attempting to re-supply and move troops to and from Guadalcanal at night by naval vessels. In an effort to stem this flow of men and equipment, American PT boats patrolled the waters off Guadalcanal to locate and intercept the Japanese vessels. In turn, Japanese float planes flew cover for their vessels and attempted to intercept the PT boats. Although developed primarily for reconnaissance over long stretches of water, these bi-planes were also employed in intruder roles.[18] When spotting an American PT boat, a float plane typically dropped flares and would then dive to attack the craft.

Detachment B personnel developed a plan to counter these Japanese tactics. If a PT boat could be utilized as bait to attract a Japanese float plane, a waiting P-70 might be able to spot and bring down the aircraft. During the evening hours of 16 August a P-70 with First Lieutenant Jack Bradner (pilot) and Flight Officer Willie Coleman (R/O) departed Munda, on New Georgia Island, in the Solomons with the mission to bag one of the Japanese float planes that harassed American PT boats. The plan included employing three PT boats to patrol the waters near New Georgia as bait in an attempt to lure a Japanese aircraft. The weather was good and the night time visibility was estimated at three miles, making it a prime evening for night hunting. During their first sally through the area the boats were unopposed. The boats turned and made a second pass while firing their machine guns at a Japanese held coastal area. This time a Japanese float plane interrupted the display of force. The Japanese pilot dropped at least one flare and dove upon the fast moving PT boats below him. Unknown to the Japanese pilot and his gunner, Bradner and Coleman lurked nearby in the darkness.

The American crew, now fully alerted to the presence of a Japanese aircraft, banked their P-70 for an attack upon the enemy war bird. Bradner squeezed off a quick burst of his guns at the bi-plane, employing a very difficult 90 degree deflection shot. When firing at an enemy aircraft, a pilot or gunner must shoot at the point in the sky where the bullets will meet a moving aircraft, rather than at the location where he sees the plane upon pulling the trigger. An exception is when the attacking aircraft is directly behind an enemy plane flying a straight path. Bradner did not score any visible hits with his wide deflection shot on the first pass. The American pilot

Flight Officer Willie Coleman of Detachment B, 6th Night Fighter Squadron. Coleman flew as the Radar Observer for First Lieutenant Jack Bradner (pilot) during a P-70 Nighthawk mission to bag Japanese float planes harassing American PT boats. The P-70 crew jumped an unsuspecting float plane that had given away its position by dropping a flare. After a brief duel in which the P-70 pilot fired upon the float plane at least twice, the Japanese plane pulled a quick maneuver and disappeared. Coleman was not able to acquire the Japanese aircraft on his radar scope. Bradner later commented, "I was sure that I got the float plane! The captain of the PT boat stated that the enemy aircraft went down, but I never received credit. It didn't really matter, because after this mission, the 'floats' didn't bother the PTs!" *Courtesy of Don Dessert.*

Front left to right: First Lieutenant Dick Stewart, First Lieutenant Bob MacDonald, First Lieutenant Don Dessert, and First Lieutenant Charles Van Bibber. Rear left to right: First Lieutenant "BB" Johnson, Captain "Doc" Lin Felder, First Lieutenant James Irwin, and First Lieutenant Bob LaCroix. Felder served as the squadron physician. All were members of the 419th NFS when the photograph was taken. Johnson and Van Bibber originally flew with Detachment B of the 6th NFS before becoming members of the 419th NFS when the units merged at the end of 1943. Van Bibber made confirmed kills of Japanese Mitsubishi A6M Zero fighters while flying P-38 Lightnings for Detachment B on June 16, 1943, and November 6, 1943. *Courtesy of Don Dessert.*

brought his P-70 around and into a tight pursuit of the floatplane. This time, he maneuvered to within 150 feet behind the aircraft and fired his guns again. However, only one of his 20 mm cannons fired. As the rounds left the P-70's belly, the Japanese plane pulled a defensive fast split-S maneuver and disappeared. Coleman was not able to acquire the Japanese aircraft on his radar scope. Bradner later commented:

"I was sure that I got the float plane! The captain of the PT boat stated that the enemy aircraft went down, but I never received credit. It didn't really matter, because after this mission, the 'floats' didn't bother the PTs!"[19]

Detachment B proved resourceful with their borrowed P-38s after commencing their re-training during the last week of May. Between mid-June and mid-September 1943, Detachment B pilots downed five Japanese bombers and one fighter and damaged one bomber in their P-38s. First Lieutenant Charles Van Bibber shot down a Mitsubishi A6M Zero fighter on 16 June. First Lieutenant Ralph Tuttle became the first Detachment B pilot to score a bomber with a P-38. Tuttle blasted the Betty bomber on 12 July. On 13 July Captain Emerson Barker, Detachment B's commanding officer, damaged a Betty bomber. Two nights later First Lieutenant James Harrell knocked down a Japanese Betty. Harrell repeated his feat on 14 August, and this was again duplicated the next night by First Lieutenant Henry Meigs.[20]

On 21 September Meigs made the most spectacular engagement during the Detachment's tenure on Guadalcanal with the scoring of his second and third kills in the P-38. He made visual contact with a group of Japanese Betty bombers over Guadalcanal that night. In less than 60 seconds, Meigs shot down two Bettys to the excitement of all those on the ground watching the engagement. At approximately 4:00 AM Meigs spotted the Betty bombers of the 702nd *Kokutai* (Naval Air Group) over Guadalcanal. He approached one Betty that was illuminated in a searchlight and exchanged fire with a gunner aboard the bomber. Meigs killed or wounded the gunner and the latter's weapon fell silent, but not before putting a hole in one of the P-38's elevators. The P-38 shuttered, but Meigs still managed to put accurate machine gun fire into the right wing root of the Betty, which then banked and went down in flames. Meigs then turned his attention to the second Betty bomber and was again struck by defensive machine gun fire. The P-38 pilot opened up with a full burst of machine gun and cannon fire in response and the Japanese plane plummeted to the ground. Major General Twining personally awarded Meigs the Silver Star for the accomplishment, which not only brought down two bombers in quick succession, but also boosted morale across the entire island as the tale of the feat spread among those who had not witnessed the night show. (Meigs was officially assigned to the 339th Fighter Squadron but was "seconded" to fly with Detachment B.)[21] Van Bibber scored the last kill for Detachment B prior to the unit's merger into the 419th NFS when he shot down his second Japanese Zero fighter on 6 November.

During the same period that Detachment B experienced some successes with the P-38s against Japanese raiders, the unit also suffered its own losses during operations against the Japanese on and around Bougainville, although each resulted from an accident rather than combat damage. On 11 September First Lieutenant William McIver survived a night take-off accident while piloting a P-70 at Munda. The aircraft was a total loss. Major Emerson Barker lost a P-38 during a night landing on 5 October at Munda which was also a total loss. Barker walked away from the accident. Detachment B lost another P-70 three days after the arrival of the 419th NFS. First Lieutenant Robert McDonnell survived a night landing accident on Munda during the night of 19 November which resulted in the complete loss of the aircraft.

A review of the 13th Air Force night fighters in the Solomons would be incomplete without mentioning the United States Navy and United States Marine Corps squadrons flying night operations in the area. The Navy and Marine Corps developed their night fighter programs separately from the Army Air Force but followed a similar pattern as their counterpart. At the opening of World War II, the Navy and Marine Corps dispatched officers to observe the night fighter program of the Royal Air Force and return with the foundational knowledge for developing their own program. Initially, the Marine Corps requested to be fielded with the newly developed P-61 Black Widow. However, delays in its production and the commitment of Northrop to fulfilling Army orders forced the Marines to search for an alternative night fighter. The Marines discussed acquiring the P-70 Nighthawk, but its inadequacies as a night fighter over Guadalcanal persuaded them to continue searching. They finally settled on the Vega Ventura, a twin engine medium bomber.[22] Modifications of the Ventura bomber for night fighting included the addition of six machine guns and a radar dish in the nose of the aircraft. The Navy, meanwhile, chose a single engine night fighter. The Vought F4U-2, the night fighter variant of the F4U Corsair, carried the APS-6 radar system in a radar dome attached to its right wing.[23]

While the mission of Detachment B involved the protection of the Army Force's bases, the primary purpose of separate Navy and Marine squadrons included the protection of naval ships and bases that primarily comprised non-Army assets. However, despite this theoretical difference, the three squadrons essentially protected the same general areas of the Solomons. According to the pilots of the 419th NFS who arrived in November 1943, there was little coordination between their unit and the Navy and Marine night fighters. The only official coordination occurred at Commander, Air, Solomon Islands (COMAIRSOLS), the organization to whom the 13th Air Force, Naval aviation, and Marine Corps aviation reported.

The Marine Corps' first night fighter squadron, VMF(N)-531, arrived in the Solomons 11 September. The five PV-1 Venturas of the squadron began conducting combat operations from their base at Banika, in the Russell islands, just northwest of Guadalcanal. Six attempts to intercept bogeys proved as fruitless as those made by Detachment B flying from Guadalcanal. Inexperience of the GCI operators resulted in inaccurate altitude information and the slow relaying of vector coordinates in many cases. By late September three of the Venturas transferred to Munda, on New Georgia, where they joined their Navy counterparts.

The Navy's VF(N)-75 arrived on New Georgia with six F4U-2 Corsairs on 23 September. Upon arrival of the Venturas of VMF(N)-531, the Navy and Marines developed a system to coordinate their efforts while maximizing the attributes of their individual aircraft. The Venturas, like the Army Air Force's P-70s, were ineffective as high altitude interceptors, so they were assigned to patrol below 15,000 feet, while the Navy's Corsairs operated above that altitude. Navy and Marine night fighters were to remain outside the anti-aircraft gun and searchlight ranges and could not take off for an intercept if a 13th Air Force P-38 was conducting a searchlight mission at their location.

The Marine and Navy night fighters assumed the mission to protect the ships assembled for the invasion of Bougainville on 1 November, while the P-38s and P-70s of Detachment B and later the 419th NFS primarily flew ground attack operations. VF(N)-75 scored the first confirmed kill for the Marines and Navy the night before the invasion. A Corsair of VF(N)-75, piloted by Lieutenant Hugh O'Neil and operating in coordination with the GCI of VMF(N)-531, intercepted and downed a Mitsubishi G4M Betty bomber off the coast of Vella

Flight Officer John Radzieta at a foxhole on Guadalcanal. *Courtesy of John Radzieta.*

Lavella. The Marines scored their first night kill two weeks later at 4:20 AM on 14 November. Captain Duane Jenkins and his crew were patrolling southwest of the Allied held area of Torokina, on Bougainville, when the fighter control on board an American cruiser reported multiple bogeys approaching the vessels. Jenkins and his crew quickly approached, spotted, and identified one of the incoming bogeys as a Betty bomber. The American pilot maneuvered behind the Betty and opened up with a four second burst of his machine guns when within 800 feet of the Japanese bomber. Immediately, the starboard nacelle of the bomber burst into flames as the plane began a shallow dive to lose its pursuer. Jenkins remained with the Betty and fired a second burst into the port wing root followed by a third burst. The Ventura's upper turret gunner delivered a two second burst into the fuselage of the bomber. The Betty went into a sharp dive trailing flames and exploded upon hitting the water. Radar on the American cruiser indicated that the other three bogeys turned and withdrew after their comrade went down in flames.[24]

The 419th NFS arrived on Guadalcanal on 16 November as the 13th Air Force's first complete night fighter squadron. By this date, Detachment B had three P-70s and eight P-38s in its inventory and was primarily flying night intruder missions around Bougainville with the Nighthawks. The air crews of the 419th NFS expected to receive P-61 Black Widows upon debarking at Guadalcanal and were demoralized to see the few war weary P-70s and P-38s of Detachment B awaiting them. Most of the six weeks between the 419th NFS' arrival and the end of the year was devoted to training missions and providing the pilots with transition orientations to the P-38. However, the 419th NFS did fly a limited number of combat missions during this six week period.

Like the air crews of Detachment B before them, the newly arrived 419th NFS pilots were forced to become night intruders due to the delayed arrival of a successor to the P-70. However, the 419th NFS pilots possessed two significant advantages compared to their compatriots of Detachment B. First, the unit included the former Royal Air Force (RAF) and Royal Canadian Air Force

(RCAF) pilots who had flown night intruder missions over Europe. Second, these European campaign vets and other cadre in Florida had trained the other pilots in the 419th NFS on night intruder tactics. Although the pilots were upset that their promised P-61 Black Widows were not awaiting them at Guadalcanal, they were at least familiar with the intruder role thrust upon them by the lackluster performance of the P-70 to intercept Japanese raiders. The Marine Corps' 3rd Division landed on the southern coast of Bougainville on 1 November, and the day and night intruder missions were flown in support of these men as they secured a toe hold on the island.

First Lieutenant Al Lukas utilized his Royal Canadian Air Force experience to help train the 419th NFS American night fighter crews with whom he deployed to Guadalcanal. He joined the 419th NFS after being confined to the barracks in Florida following an outing of buzzing topless co-eds sunning on dorm roofs and landing a small plane on a golf course for a drink at the club bar. After forgetting to turn off the light in his quarters, Lukas took his pistol and simply shot it out. The military offered Lukas the choice between a combat tour with the 419th NFS or a court martial. Lukas wrote in his flight log, "I decided to go overseas again. Can't stand the war in Florida." *Courtesy of Leslee Lukas Mays.*

Former RAF pilots like Al Lukas were well experienced in night intruder operations and continued to pass that knowledge to squadron mates as the 419th NFS assumed the night intruder mission of Detachment B. Perhaps some of the more important lessons learned by Lukas during his combat experience with the RCAF's Number 418 Squadron included survival and return after suffering combat damage. While every combat mission is different, stories told by Lukas and others provided comfort and sometimes awe among the other pilots. If the European campaign vets could hit their targets and survive combat damage to their aircraft, so could the other pilots of the 419th NFS.

Lukas frequently related one story to his fellow pilots which occurred on April 1, 1942—definitely a bad "April Fool's Day" for the Chicagoan. Lukas and his crew were aboard his Boston bomber *Yankee Doodle* and were "stooging" (British night fighter terminology for a plane waiting for a target of opportunity) near the German airfield at Brussels, Belgium. Anti-aircraft batteries discovered the Americans and hurled a barrage skyward at them. The *Yankee Doodle* sustained several flak hits that damaged the electrical system, a fuel line, and the internal communications. Unable to converse with his navigator, Lukas quickly departed the area and flew toward Lille, France, where he dropped his bomb load on a railroad marshalling yard before turning west toward England and safety. He arrived at the English coast near the mouth of the Thames River, a considerable distance from his home base at Bradwell Bay. Damaged and without navigational aids, Lukas flew along the Thames River in search of a roost before the British anti-aircraft guns mistook him for a German night fighter or he encountered the barrage balloons outside of London.

The British developed an emergency system for its lost and/or damaged night fighter crews. Various air bases around eastern and southern England were designated for emergency night landings. Each night, the RAF selected a different group of bases to fulfill the role and illuminated a beacon located near each installation. A night fighter pilot seeking a darkened air base could approach the beacon at an altitude below 500 feet. Arriving at the beacon, the crews would complete a series of either clockwise or counterclockwise turns around it, normally ranging from one to three in number. The British altered the maneuver each night. Upon successful completion of that night's pattern, a nearby RAF base would illuminate its runway lights for an emergency landing. An unsuccessful attempt to complete the flying pattern would be met with a hail of anti-aircraft fire. Lukas spotted one of the emergency beacons and flew toward it. Upon the completion of the night's circular pattern, Lukas was rewarded with the welcoming lights of an RAF airfield. He related to the author later that he did not care if he was up or down wind—he planned to put his Boston III on the ground. As the *Yankee Doodle* raced down the runway following touchdown, both engines died due to the loss of fuel. A slightly larger hole in the fuel line would have left the crew parachuting for their lives over the English Channel or along the Thames River.

Lukas braked the plane and all three men scrambled from the aircraft. He described the moment as one of bear hugs and "banshee" yells as they celebrated their safe arrival. Emergency vehicles arrived on the scene and drove the three Americans to a mess hall where hot tea awaited. Lukas learned that he had arrived at West Malling. Ground crews repaired the damage and the trio flew their Boston III on a 50 minute flight back home to Bradwell Bay.[25] Despite any comfort provided by the knowledge that a damaged night intruder pilot could bring his crew home from enemy territory, the 419th NFS personnel were not prepared for their first casualty of the war only a week after arriving on Guadalcanal.

419th NFS Commander Captain John McCloskey was flying a routine P-70 intercept training mission during the night of 22 November when something tragically went wrong. The mission that night involved practice intercepts of an accompanying P-38 flown by First Lieutenant Don Dessert. At the conclusion of the training Dessert landed his P-38 and awaited McCloskey. As McCloskey and his crew, Second Lieutenant John Blankenship (R/O) and First Lieutenant George Budway (assistant intelligence officer), approached Guadalcanal to land, the P-70 unexplainably went down into the water approximately 100 yards from the shore at Kukuma Beach. The plane inverted itself as it came to rest partially submerged closer to the shore. 419th NFS personnel on the ground watched in horror and then rushed to the scene. Dessert, First Lieutenant Dick Stewart, and Corporal Theodore Knutson swam out to the wreck to save the crew who were trapped in the P-70. Fuel poured into the surrounding water as Dessert and Stewart pulled Blankenship and Budway from the wreck. However,

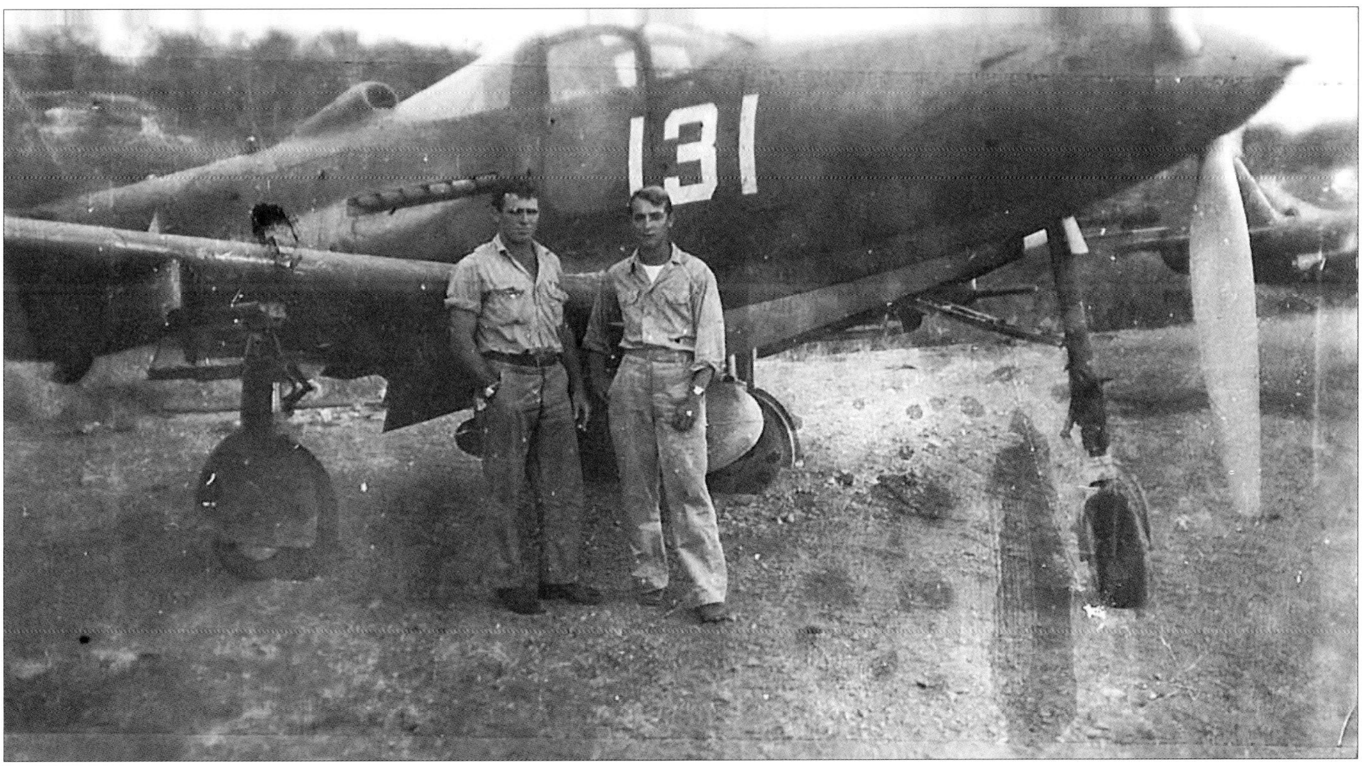

Left to right: Second Lieutenants Don Dessert and Dick Stewart stand in front of another squadron's Bell P-39 Airacobra on Guadalcanal. Dessert and Stewart often borrowed a P-39 for hops in the area. Both noted that the plane seemed too heavy compared to the more nimble P-38 Lightning. *Courtesy of Don Dessert.*

Kneeling left to right: Second Lieutenant Dick Stewart, Second Lieutenant Don Dessert, Second Lieutenant Herman Manke, and Second Lieutenant Bob MacDonald. Standing left to right: Second Lieutenant Bob LaCroix, Second Lieutenant Bill Mackie, First Lieutenant James Irwin, and First Lieutenant "BB" Johnson. The officers are posing in front of P-38 # 301 *99 Bottles*. *Courtesy of Dick Stewart.*

they were not able to reach McCloskey, who died within the inverted plane. The fuel in the water caused severe but temporary irritation in the eyes of the rescuers as they struggled from the water with Blankenship and Budway.

Stewart remembered that they managed to get a sling over the P-70 and lift it at a 60 degree angle. Stewart sadly noted that he could see that McCloskey was dead within his cockpit. He pulled McCloskey from the cockpit and swam back to shore with his left hand while holding onto his deceased commander with his right hand. The effort caused Stewart to continually drop below the surface, forcing him to rise again gasping for air. As he approached the shore others on the beach helped pull Stewart from the water. Squadron Flight Surgeon Captain Lin Felder informed Stewart that he could let go of McCloskey's body. Stewart stammered, "I can't." He was so tired and racked with the shock of the event Stewart's hand would not voluntarily release its grip of McCloskey. He later remembered that it was as though he had glue in his hand and he could not open it to release McCloskey. They placed McCloskey's body inside an ambulance and Stewart was moved into the Flight Surgeon's vehicle. Stewart was transported back to his quarters and immediately fell into a deep sleep.[26] McCloskey was buried on Guadalcanal, and three P-38s flew a "missing man" formation overhead to honor the 419th NFS commander. Knutson later received the Soldier's Medal for the rescue of the two crewmen that night. Being a medal designated for enlisted soldiers only, Dessert and Stewart were not eligible to receive the decoration for their efforts to save their comrades.

The death of McCloskey was a severe blow to the morale of the entire unit. Not only was this the first fatality of the unit overseas, but McCloskey was an extremely popular and charismatic squadron commander. It was a bitter blow for the personnel of the 419th NFS to see the RCAF veteran die in a training accident days after arriving in Guadalcanal. With the arrival of the 419th NFS, pending transfer of Detachment B, and the death of the man who was to lead the night fighter assets of the 13th Air Force, an important decision had to be made by 13th Fighter Command. What should be the future of the Command's night fighters? Despite the desires of the 419th NFS for a commander to be selected from among its ranks, the decision was made to appoint Major Emerson Barker of Detachment B as the new commander of the 419th NFS. At the same time, many of the personnel in Detachment B were transferred into the 419th NFS. Others were assigned to day fighter squadrons on Guadalcanal, and some rejoined a newly reconstituted 6th NFS that was formed from the assets of Detachments A and B, as well as the squadron headquarters in Hawaii. Thus, the 419th NFS became a hybrid of the unit that arrived on Guadalcanal on 16 November and some of the personnel originally assigned to Detachment B with a squadron commander from the latter unit. Officially Detachment B merged into the 419th NFS on 15 December. The 419th NFS also inherited all of the aircraft of Detachment B and would thus continue operations flying war weary P-70s and P-38s.

The night fighters lost a P-38 during the night of 3 December. Captain Ralph Tuttle had taken off from Guadalcanal and was over the water when his engine failed. Tuttle bailed out of the P-38 but lost consciousness in the process. Fortunately, he regained consciousness just prior to hitting the water in his parachute. Search crews found him floating in his small survival dinghy.

Food quality varied greatly from Guadalcanal to the other islands of the Solomons. First Lieutenant Richard "Dick" Stewart offered one word to describe the food they received—"lousy." Sergeant Richard Pitman commented that the Navy had the best food, including plenty of ice cream, in the Solomon Islands. At the same time, he realized that the 419th NFS had better food than many other military personnel. When serving on Bougainville, Pitman noted that he would give his chocolate bars to the Army and Marine infantrymen holding the perimeter around the airfields:

"I felt sorry for them living out there in holes while protecting us. We had it so much better than them and I wanted to do something to help them."[27]

At times, 419th NFS personnel assigned to more remote areas ate the infamous K-rations. These were tactical field rations packaged in a small box for transportability. Each contained a canned meat or cheese product, crackers, candy, cigarettes, powdered drink mixes, and other items depending upon whether they

"The Spider's Web" – the 419th NFS officers' club on Guadalcanal. *Courtesy of Leslee Lukas Mays.*

were packaged for breakfast, lunch, or supper. The 419th NFS did have their own internal mess section, so the unit as a whole normally ate hot meals, although the quality was frequently questionable. Normally, the men ate canned foods which were more easily transported from the United States. Many veterans of the unit complained about having to eat canned vegetables during World War II since they were accustomed to fresh vegetables back home. Today, it is much more common for Americans to each canned vegetables at home. Breakfast normally consisted of powdered eggs and/or pancakes with coffee. True to the Dr. Seuss children's book *Green Eggs and Ham*, the powdered eggs had a green tint to them according to Stewart. Australian mutton was normally the only fresh meat the men of the 419th NFS consumed during the Solomons Campaign. Stewart remarked:

"You could smell the mutton cooking in our alert shack, which was half a mile away from the mess hall. It was cooked in a single large pot and had a couple inches of grease floating on top."

The air crews longed for their infrequent combat rest leaves in Australia and New Zealand where they could acquire fresh food. Stewart remembered desiring the fresh milk, steak, and fresh eggs, fresh fruit, and fresh vegetables available when on combat leave. Later in the war the quality of the food available to the mess hall improved significantly. In addition, the squadron would take up a collection of money from everyone and would occasionally fly their small cargo plane to Australia, where they would privately purchase fresh food for the unit.[28]

Fresh meat, other than mutton, was rare, and everyone in the unit seemed to dislike the canned meat product known as spam that was served frequently in the mess hall. After the war, Captain Lukas would not allow spam to be brought into his house. However, there is usually one person who defies group consensus, and that individual was Staff Sergeant Elzie "Eddy" Ballard. In an act that often defied all logic to his squadron mates, Ballard traded his beer rations for cans of spam. "I still love it," he explained one afternoon to the author.[29] Several of the squadron members made multiple trips to Cape Esperance, at the end of the island, to trade detested cans of spam for fresh fruit with the local inhabitants. During one trip Flight Officer Sanford "Sandy" Bauman and others "borrowed" a command car and crossed many lightly flowing streams as they drove to Cape Esperance. After the trading session the men loaded their fruit into the command car for the return journey. A rain storm in the mountains released a tremendous increase of water into the nearly dry streams the men crossed on their way to Cape Esperance. The streams began to swell into rivers, but Bauman and the others continued their journey, firmly

First Lieutenant Robert Baumgardner, the 419th NFS' adjutant. Although he insisted on following the sluggish "official" Army supply requisition system, non-commissioned officers quickly learned the art of trading and "midnight requisitions." Some resourceful squadron members made their own Japanese flags for trading with unsuspecting sailors for items needed by the squadron. *Courtesy of Don Dessert.*

Second Lieutenant Don Dessert flew the 419th NFS' first attempt to intercept a Japanese aircraft on December 10, 1943. He conducted the unsuccessful mission in a Lockheed P-38 Lightning. *Courtesy of Don Dessert.*

believing they could always use the winch on the front of the command car to pull the vehicle out of any stream in which it became stuck. However, they drove into one stream and misjudged its depth. The car and its contents were picked up by the current and swept downstream. The men managed to scamper out and make it to the other bank. Bauman noted:

"Here we were some 20 plus miles from our base with nothing to get us back over the streams. We walked to shore on the beach and spotted a Navy barge headed south. They saw our waving shirts and picked us up. When we returned, we destroyed the records of the command car at the motor pool and saved the information for this sage."[30]

The Army supply system on Guadalcanal and throughout the Solomon Islands can be described as "sluggish" at best. Sergeant Cleveland "Cleve" Wilborn, the Transportation Non Commissioned Officer (NCO), attempted to work within the existing military supply system when the 419th NFS arrived on Guadalcanal, but quickly learned that he needed to operate within an alternative system. Wilborn realized that scrounging and trading were sometimes the best methods of supply acquisition in a system full of shortages and delays. At one point Wilborn acquired a new jeep for Captain Howard Daniel and later a new weapons carrier for the squadron. This often put Wilborn at odds with First Lieutenant Robert Baumgardner, the squadron adjutant, who demanded that the transportation NCO acquire everything through the official requisition process. Other members of the 419th NFS learned scrounging and trading techniques. Air crews were permitted brief leaves in Australia or New Zealand for rest from their combat missions. They would return with as many bottles of brandy and other alcohol as possible for trading purposes. Several resourceful members of the squadron developed a war souvenir industry. They would hand manufacture Japanese flags and other items to trade with unsuspecting sailors and others passing through the area. In the Admiralty Islands, the squadron traded P-61 flights for fresh meat from the Seabees. First Lieutenant Robert "Bob" O'Brien, the 419th NFS Ordnance Officer, noted that if you could not trade for an item, one had to resort to other means to keep the squadron functioning:

"Moonlight requisitions [i.e. clandestine 'borrowing' when no one is looking] were the order of the day to get what you needed and couldn't get through channels."[31]

The 419th NFS commenced combat operations on 10 December. First Lieutenant Don Dessert flew the unit's

Left to right: Sergeant Willard "Wee Willie" Fellows and Second Lieutenant Don Dessert. Fellows served as the crew chief for Dessert. The pair is standing in front of P-38 #303 *Whoosh*, Dessert's regular aircraft and the one under the care of Fellows. One of the 339th Fighter Squadron pilots who participated in the operation that downed Japanese Admiral Isoroku Yamomato's Mitsubishi G4M Betty bomber flew *Whoosh* on that April 18, 1943, mission. *Courtesy of Don Dessert.*

first aircraft intercept mission on that day, while First Lieutenant Lukas completed the 419th NFS' first night intruder mission that same night. Dessert, flying from Munda, scrambled in search of five bogeys. However, friendly anti-aircraft fire placed a few holes in his P-38. The GCI operator ordered him to land. The bogeys returned to the area, but they did not drop any bombs on the American facilities at Munda.[32]

First Lieutenant Lukas flew the 419th NFS' first night intruder mission on 10 December at approximately the same time Dessert was chasing the bogeys near Munda. Lukas, along with Second Lieutenant John Blankenship (R/O) and Staff Sergeant Glen "Frosty" DeForrest (gunner), targeted the Japanese supply base of Bonis, on Bougainville. The P-70 Nighthawk lacked an internal bomb bay because that area of the plane carried a belly tub with guns for engaging Japanese aircraft at night. The P-70s of the 419th NFS modified their few P-70s to carry a small bomb load for night intruder missions. First Lieutenant O'Brien noted:

"The P-70s had no equipment such as bomb racks to load bombs on. Our Armament Officer [Second Lieutenant Clarence 'Tom' Erb] and his men scrounged these items from B-25s in the local salvage yard and installed them along with controls to jettison them. That is why we were able to fly intruder bombing missions. We also tried installing rockets on the wings. We installed two on a P-70, one on each wing, and fired them on a practice mission. The pilot used a grease pencil to make an 'X' on the windshield as his aiming circle. I flew in the back on that mission, and we thought we really had something, but had to terminate it since we could not get any more rockets. We were not authorized them and could not find a place to 'midnight requisition' them."[33]

First Lieutenant Al Lukas flew the 419th NFS' first P-70 Nighthawk intruder mission on December 10, 1943. Lukas announced his evening arrival at Japanese held Bonis airfield, on Bougainville, by dropping four 100 pound high explosive bombs from an altitude of 800 feet. *Courtesy of Leslee Lukas Mays.*

When asked to explain the differences in duties of the armament and ordnance officers O'Brien offered:

"The Armament Officer is responsible for loading the armament aboard the aircraft (bombs, rockets, machine gun ammo, etc.), while the Ordnance Officer orders, stores, and delivers these items to the Armament Officer at the plane and repairs the weapons when needed. They work as closely as a husband and wife."[34]

As a veteran of 19 intruder missions over German occupied Europe, Lukas repeated many of the lessons he learned from that theater of operations. The crew took off in their P-70 from Munda at 7:15 PM and proceeded on a northerly course toward Bougainville. Reaching the northeast coast of the island, Lukas located the Japanese Buka airfield and orbited nearby to prepare his attack on the Bonis supply base. Banking left, the pilot made his run on Bonis, located southeast of Buka. Lukas reached Bonis at 9:20 PM and announced

Left to right: First Lieutenant Dick Stewart, First Lieutenant James Irwin, Second Lieutenant Bob LaCroix, and Second Lieutenant Herman Manke posing on a bomb. *Courtesy of Dick Stewart.*

his arrival by dropping four 100 pound high explosive bombs from an altitude of 800 feet. DeForrest reported that he saw each bomb explode. A Japanese anti-aircraft gun opened up on the P-70 from the northeast corner of the base, and it was answered by DeForrest with his machine gun. Lukas wheeled the P-70 around and made a second run across the base. With their bomb load expended, the crew utilized this fly by to strafe the Japanese facilities below them. Lukas fired the forward .50 caliber machine guns while DeForrest continued to fire his gun at the rear of the aircraft. Lukas knew from his European experience that the enemy were now prepared for a third attack across Bonis, so he departed the area rather than risk the anti-aircraft guns that would be manned and ready for his return. Lukas and his crew returned to Munda at 11:55 PM after a successful night intruder mission.[35]

The next day the 419th NFS lost two P-38s in separate accidents. Dick Stewart crashed during an attempted take off in one of the P-38s. Stewart noted that the P-38's engine simply quit operating during the take off. He plowed through a group of palm tree stumps at the end of the runway before the aircraft came to rest on its nose. Stewart managed to walk away from the P-38 without injury, but the plane had to be written off as a total loss. Mechanics determined later that the P-38 suffered a compressor failure. If the plane had managed to get partially airborne before losing engine power the situation could have been serious for Stewart.[36] On the same day, First Lieutenant Charles Hicks totaled another P-38 during a day landing attempt. Hicks, another of the RCAF vets of the 419th NFS, cracked up during his initial flight in a P-38. Second Lieutenant John Radzieta, who trained with Hicks in Florida and served as the pilot's R/O on the P-70, remembered:

"Hicks made a hot landing. The braking system of the P-38 was different than the one in the P-70. In the P-38 you had to pump the brakes to build up pressure. In the '70 all you did was hold them down. Before he could stop fully or turn off onto the taxi strip he spotted a fuel truck in the middle of the taxi strip at the edge of the runway. He elected to roll beyond the end of the runway. The plane rolled into a bomb crater filled with water and flipped end over end and landed on its back. He could not open the hatch. Hicks crawled out through one of the side windows, which was almost an impossible feat. He decided that he wanted no part of the P-38. Hicks was reassigned as a liaison officer between the Army Air Force and the Navy."[37]

Pilots who decline to fly for any reason are reassigned according to the needs of the military. Hicks was a well-respected and liked pilot and was missed by all when he left the unit.

The newly arrived 419th NFS discovered that flying around Guadalcanal at night could be hazardous for many reasons, including friendly anti-aircraft (AA) gunners mistaking the night fighters as Japanese raiders. Flight Officer Sandy Bauman, an R/O, noted that he was flying his first operational flight with First Lieutenant Herman Manke, pilot, when they accidentally flew too far to the northeast and approached the naval base on Tulagi. Searchlights swung into action and illuminated the P-70 and its startled crew. Night regulations prohibited the crew from contacting the Navy to inform them that they were friendly, so they remained at the mercy of the anti-aircraft gunners aboard the vessels. "I told Herman that at the first sign of anti-aircraft fire, I would bail out and he agreed also," recounted Bauman. "They did not fire and we returned scared as hell, but lucky."[38]

The 419th NFS lost its first aircraft and crew during a combat mission on 14 December. Following the squadron's first intruder mission on 10 December, pilots of the 419th NFS flew three more intruder operations during the next three days. Their fifth intruder mission on 14 December proved fatal for three crew members. Second Lieutenant Dean Vanderhoff (pilot), along with Flight Officer Vincent Mays (R/O) and Staff Sergeant Council Roberts (gunner), were conducting a night intruder mission over Bougainville when they were hit by friendly fire. Marine anti-aircraft gunners at Torokina air strip spotted the P-70 and mistook it for a Japanese bomber. They opened fire and struck the aircraft, which then crashed in a swampy area near the American air strip. Initially, the Marines claimed the destruction of a Japanese bomber and Vanderhoff was reported missing and presumed shot down by the Japanese.

It was not until March 1944 that the facts became known in the incident. Following the defeat of a major Japanese counter attack in March (related in Chapter 3) an American patrol came across the downed P-70. The wreckage consisted of many small pieces of the plane scattered across the swamp. The largest piece held the tail number of the aircraft, verifying that it was Vanderhoff's P-70. The 419th NFS formed its own accident board and investigated the crash site. The only human remains ever recovered were the skeletal foot of the R/O, which was found still inside his boot. The incident proved to be a terrible blow to the personnel of the 419th NFS—their first combat related fatalities had been as a result of friendly fire.[39]

Second Lieutenant Herman Manke. The pilot strayed off course while patrolling near Guadalcanal with Radar Observer Sandy Bauman in a P-70 Nighthawk. As they approached the American naval base on Tulagi, searchlights swung into action and illuminated the P-70 and its startled crew. Night regulations prohibited the crew from contacting the Navy to inform them that they were friendly so they remained at the mercy of the anti-aircraft gunners aboard the vessels. "I told Herman that at the first sign of anti-aircraft fire, I would bail out and he agreed also," recounted Bauman. "They did not fire and we returned scared as hell." *Courtesy of Don Dessert.*

Second Lieutenant Dean Vanderhoff. On December 14, 1943, Vanderhoff piloted a P-70 Nighthawk on an intruder mission over Bougainville with Flight Officer Vincent Mays (radar observer) and Staff Sergeant Council Roberts (gunner) when they were hit by friendly fire. Marine anti-aircraft gunners at Torokina air strip spotted the P-70 and mistook it for a Japanese bomber. The Marines opened fire and struck the aircraft, which then crashed in a swampy area near the American air strip. *Courtesy of Leslee Lukas Mays.*

Left to right: First Lieutenant Dick Stewart, First Lieutenant Don Dessert, First Lieutenant "Marvelous" Marvin Walker, and First Lieutenant Charles Eckstein, who transferred to the 419th NFS when the unit absorbed Detachment B, 6th NFS. *Courtesy of Don Dessert.*

An approaching Japanese Sally bomber on 23 December offered the 419th NFS an opportunity to intercept it with P-38s. However, rather than downing the bomber, the 419th NFS pilots seemed to be plagued with countless problems that morning. Dick Stewart and First Lieutenant Don Gardner were airborne at the time. Stewart flew at a lower altitude, while Gardner patrolled much higher over Guadalcanal. Stewart saw the bomber, but was so far below it he could not gain sufficient altitude before it had left the area. Gardner was at an excellent altitude for catching the Sally, but he was experiencing radio failure at the time and could not be called and informed of the Japanese bomber's presence. First Lieutenant Marvin Walker attempted to take off and support the other two pilots in a pursuit of the Sally. As Walker darted down the runway to take off the P-38 lost an engine. Walker managed to bring the P-38 to a halt without cracking up at the end of the field. In the meantime, the Japanese bomber safely slipped away.[40]

Lukas made his second night intruder mission of 1943 on 26 December. This time he flew with Flight Officer Lane Thompson (R/O) and Corporal M. K. Riddenhour (gunner). Night intruder crews did not attack their targets at the same time each night they were on the prowl. To do so would alert the Japanese when to be prepared for the arrival of the P-70s. On this mission, Lukas was airborne at 2:40 AM from Munda and destined for the area around Bonis, on Bougainville. Lukas approached Bougainville at 5,000 feet and dropped to 800 feet as he neared the island. From 4:00 AM to 5:00 AM Lukas flew back and forth between Tinputs Harbor and Hangan to confuse and disturb the Japanese below as to where he might be destined as they heard the P-70's engines. At times he turned inland to add to their confusion. Japanese aircraft never challenged the P-70 due to the darkness

Front left to right: Flight Officer William "Willy" Coleman and Flight Officer John Lomell. Rear left to right: First Lieutenant Ken Voy, Flight Officer Ralph Lillard, and Flight Officer George Black. *Courtesy of Don Dessert.*

and the fact that Allied attacks had dramatically reduced Japanese airpower in the immediate area. Between 5:00 and 5:30 AM Lukas made a dozen passes over known Japanese supply and bivouac areas, dropping two 100 pound bombs on Bonis Airfield. During strafing runs Lukas expended 2,000 rounds of .50 caliber ammunition on Japanese facilities, while Riddenhour put 100 rounds into a radio tower. At 5:30 AM Lukas departed the area and flew over Chabai Plantation and turned toward Baniu Harbor, where he observed lights apparently coming from a vehicle parked under some trees. He turned his P-70 and strafed the lights, noting that they went out during the attack, although he could not be sure if this was because of bullet strikes or if an operator turned them off to darken the target. The P-70 crew continued down the southeast coast of Bougainville at 700 feet and landed at Munda by 7:00 AM.[41]

Stewart and First Lieutenant Dorval Brown surprised a Japanese patrol on Bougainville and killed an estimated 30 soldiers on 30 December. Stewart was flying his first combat patrol in the P-38 after sitting at Torokina air strip, on Bougainville, as the night alert plane. Stewart and Brown were making the routine daylight patrol when the former spotted movement along the beach and signaled his wingman. Closer inspection revealed a large Japanese column walking along the surf and unaware of the danger above them. Stewart dropped his wing tanks prior to engaging the Japanese force as instructed in pilot school back in the United States. Brown did not release his wing tanks to the amazement of Stewart, who wondered at the time if this could be a battlefield modification to "Stateside" training.

Stewart lowered the nose of the plane and commenced his dive on the soldiers below. As he approached the column, the pilot could see two Japanese soldiers carrying a machine gun on a pole and an officer walking with them, while the other soldiers trailed behind the three men. Stewart carefully placed the three men in his gun sight and fired as he barreled down upon the beach. The

The 419th NFS utilized this Douglas A-24 Banshee (the Army version of the Navy's SBD Dauntless) dive bomber as an instrument trainer. *Courtesy of Don Dessert.*

Japanese soldiers jerked and twisted before dropping into the sand as Stewart desperately yanked on his stick to avoid slamming into the beach. Having concentrated so intently on lining up and firing his guns, Stewart had failed to keep track of his speed and diving angle! Safely pulling out of the diving attack, Stewart glanced back to watch Brown complete his run across the beach. Brown finally remembered to drop his wing tanks as he approached the column. However, he forgot to turn on his cross feed so that fuel would flow from the main tank. Both engines died as the fuel tanks fell from the wings. Brown passed over the startled Japanese column while desperately struggling to initiate his fuel flow from the main tanks. The P-38's engines roared to life as the relieved pilot elevated the plane's nose and cleared the area.

Stewart banked his aircraft and returned for a second pass on the Japanese soldiers, who by this time had scrambled into the nearby tree line. The pilot directed his machine gun fire into the trees and returned several times with Brown. When they had exhausted their ammunition the pilots departed the area and returned to Torokina air strip. Stewart's crew chief approached the young pilot after he landed and asked, "What happened, Lieutenant"? Stewart related the strafing runs along the beach, but seemed confused by the crew chief's concern. The sergeant thumbed toward the tail of the P-38 and quizzed, "Have you seen the back end of your airplane"? Stewart walked to the rear of the P-38 to discover the plane peppered with bullet holes. A count resulted in a tally of 89 holes—each having missed a vital component

Second Lieutenant Dick Stewart. *Courtesy of Dick Stewart.*

In the Jack Fellows print "Surprised," Second Lieutenant Dick Stewart machine guns a Japanese patrol on December 30, 1943, as First Lieutenant Dorval Brown pulls up and away from the Bougainville beach in his Lockheed P-38 Lightning. *Courtesy of Jack Fellows.*

of the aircraft. Stewart confided that this was the one time in his combat service that he became physically ill and began vomiting after realizing how close he came to becoming a casualty or being shot down.

There is a postscript to this combat mission. The military admitted Stewart, suffering from a severe case of dysentery, to the Bougainville military hospital in early 1944. He was assigned to a bed next to a New Zealand Major assigned to the infantry battalion from Fiji located on the island. The Fijian battalion developed a reputation for long distance reconnaissance and raids against Japanese installations across the island. The Major greeted Stewart and asked about his assignment to Bougainville. Stewart answered and recounted his experiences, including the attack along the beach. The New Zealand officer's eyes widened, and he asked Stewart how many Japanese soldiers he thought died in the clash. The pilot replied that he estimated between 10 and 20 Japanese soldiers. The Major vehemently shook his head and explained that the tally was much higher. Several days after the incident, the Major and a Fijian patrol came across the bodies of the Japanese soldiers killed in the attack on the beach. He explained that approximately 30 Japanese soldiers perished as a result of the strafing runs that day. The officer commented that he retrieved the Japanese officer's sword. The Major requested one of his Fijian soldiers to deliver the sword to the hospital and then presented it to Stewart, who brought it home after the war.[42]

On the last day of 1943, First Lieutenant Donald Gardner became the first pilot of the 419th NFS to be shot down by the Japanese. On 31 December Gardner was downed by Japanese anti-aircraft fire while on a daylight strafing mission in a P-38 over Kahili airdrome, on Bougainville. Gardner, flying with Henry Meigs, the former Detachment B pilot, strafed the airfield. Both men placed machine gun bullets into a Betty bomber as they passed and observed it to be burning as they pulled up from their attack. At this point Japanese anti-aircraft fire hit Gardner's left engine. Gardner managed to coax his P-38 out to the sea, where it would be safer for him to parachute from the stricken plane. He bailed out from his plane and Meigs made an emergency call over the radio for a rescue attempt. Meigs continued to circle near

Standing left to right: First Lieutenant Bob Oates, First Lieutenant Al Lukas, First Lieutenant Bob Winslow, and First Lieutenant John Laffey. Kneeling left to right: Unknown, First Lieutenant Dick Stewart, Second Lieutenant Herman Manke, and Corporal Carl Ranck. *Courtesy of Leslee Lukas Mays.*

419th NFS air crews pose on a P-38 Lightning on Guadalcanal. Note First Lieutenant Dick Stewart holding the squadron mascot, *Old Salty Dog*, on the nose of the aircraft. *Courtesy of Leslee Lukas Mays.*

Left to right: An unknown crew chief and Second Lieutenant Bill Mackie. Both men are standing in front of the P-38 #304 nicknamed *Midnight Marjorie* by First Lieutenant Al Lukas for his wife (the former Marjorie Kingsley), a well-known singer with Harry Roy and his Orchestra in England. *Courtesy of Don Dessert.*

his downed comrade, who was successfully rescued by a Navy Consolidated PBY Catalina only 20 minutes after hitting the water.[43] 1943 ended ominously for the night fighters of the 13th Air Force. The 419th NFS arrived to serve as the sole night fighter squadron of the 13th Air Force and relieve Detachment B. However, between 11 September and 31 December Detachment B 6th NFS and the 419th NFS lost four P-70s and five P-38s to accidents and combat while not downing any Japanese aircraft. Four air crew members who arrived with the 419th NFS on 16 November died during the short six week period ending on 31 December. Night fighting was not easy, and it would not become easier in 1944.

3

THE SOLOMON ISLANDS (1944)

"We lost Bob MacDonald flying my wing to heavy anti-aircraft fire. He had a good parachute into the jungle where coast watchers confirmed that he shot it out with a Japanese patrol, resulting in his death."

First Lieutenant Don Dessert
Pilot, 419th NFS

1944 opened with Allied forces holding onto a western corner of Bougainville for a second month as 5th and 13th Air Force aircraft continued to hammer at Japanese air bases in the area. Since October 1943, the 5th Air Force had assumed primary responsibility for air attacks against the Japanese bases around Rabaul, while Commander Air Solomons (COMAIRSOLS) joint command, including the 13th Air Force, concentrated on Bougainville and the surrounding islands. The Japanese utilized Rabaul, on New Britain, as their major base in support of operations in New Guinea and the Solomon Islands. Four airfields with revetments for 166 bombers and 265 fighters surrounded Rabaul, while a fifth fighter base lay across the Saint Georges channel on New Ireland. Rabaul's Simpson Harbor served as a naval base, permitting the arrival of men and materiel to support the Japanese war effort in the Southwest and South Pacific.[1]

The arrangement between the 5th and 13th Air Forces permitted the South Pacific forces of Admiral William F. "Bull" Halsey to land on Bougainville and establish a toehold on the island in November 1943, while the air power of General Douglas MacArthur's Southwest Pacific forces worked to pin Japanese Army and Navy aircraft around Rabaul. In turn, during December the 13th Air Force, along with its Navy, Marine Corps, and New Zealand counterparts, assumed responsibility for the air offensive against Rabaul as the 5th Air Force supported General MacArthur's operations on New Guinea and the western end of New Britain. During December, the 13th Air Force flew daylight heavy bomber raids over targets surrounding Rabaul. Difficulty in hitting many of the targets and effective Japanese defenses persuaded General Hubert Harmon, the 13th Air Force commander, to switch to night bombing beginning 10 January. However, the 419th NFS beat the heavy bombers to Rabaul when Captain Al Lukas flew the first solo night intruder raid on 9 January.

Lukas related that the idea for the raid originated in a mess hall on Guadalcanal when he heard several senior United States Marine Corps officers discussing the difficulty in accurately bombing Rabaul. The Japanese base was situated on a harbor that had been carved over time from the crater of an extinct volcano on the northeast end of New Britain Island, located at the northwest end of the Solomon Islands chain. The terrain helped to protect the base and its harbor, while heavy Japanese air defenses, including fighters and ample anti-aircraft guns, forced Allied bombers to attempt high altitude missions.

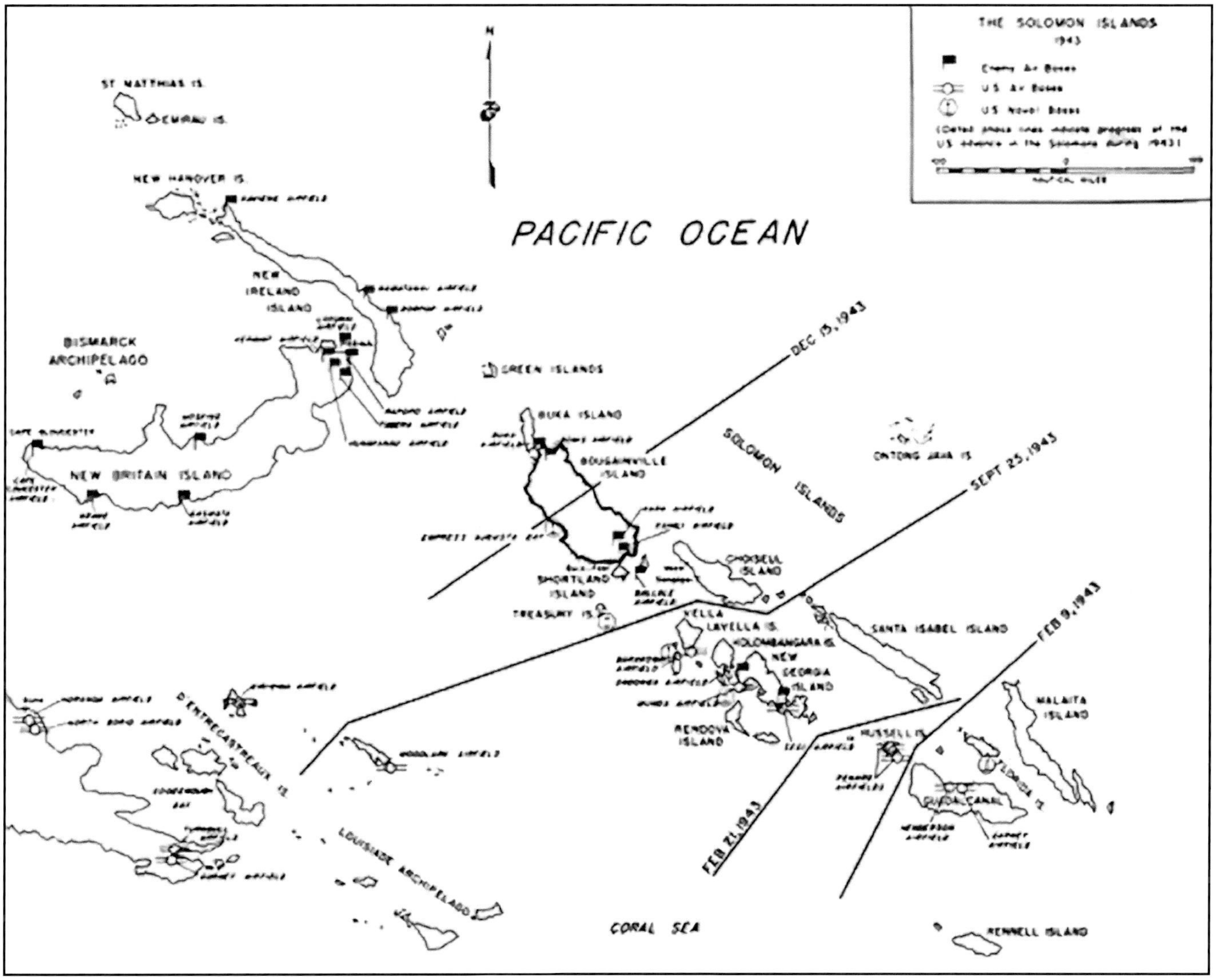

The Solomon Islands area. *Courtesy of the Center for Military History.*

Rabaul served as the major Japanese army and navy base during the Solomon Islands campaign and was a tempting target. Lukas boasted that he knew a way to accurately bomb Rabaul and immediately encountered skepticism from the officers. However, he provided details of his service with the Royal Air Force (RAF) and the low-level night bombing tactics he had learned while flying against German targets. In the European Theater he flew solo raids against German airfields, factories, and trains utilizing the Boston bomber, a version of the American A-20 Havoc. The P-70 Nighthawk was also a variant of the A-20. Thus, flying a plane that was similar to the one he piloted against the Germans, Lukas felt confident that he could duplicate his raids against German facilities with an attack on Rabaul. Although not fully convinced, the officers passed the idea forward and Lukas received permission to attempt a one-plane night raid on Rabaul.

Lukas asked his regular P-70 crew if they would volunteer to accompany him on the mission and both men accepted. This was a relief, because he commented years later that he felt comfortable with his regular crew and did not desire to substitute either due to the complexity of the mission. Second Lieutenant John Blankenship would serve as the Radar Observer (R/O), while Staff Sergeant Glen DeForrest agreed to man the guns of the P-70. Lukas poured over the intelligence photos of Rabaul to study the target and the best routes in and out of the area. Since he could not keep the photographs, he had to select the routes and targets and then commit them to memory. At Rabaul, Lukas noticed the naval docks and the fuel dump. Directly in line with the two former targets was the Australia House, which housed officers. Lukas selected these three targets due to their importance and close proximity in a straight

Captain Al Lukas (standing at right) receiving the Air Medal from Major General Hubert Harmon. Lukas flew the first solo night intruder raid against Japanese held Rabaul on January 9, 1944. Utilizing night intruder tactics he learned with the Royal Canadian Air Force, Lukas dropped his load of high explosive and incendiary bombs across a dock and a building housing Japanese officers. Upon returning to Guadalcanal, Captain Warren Slater, the 419th NFS intelligence officer, exclaimed, "That was super, Luke," leading to the life-long nickname of Al "Super Luke" Lukas. *Courtesy of Leslee Lukas Mays.*

line from his approach route across the harbor. If Allied intelligence was correct, a single bomb load could be dropped across the naval dock, fuel dump, and then into the nearby Australia House. After settling on the targets, Lukas calculated his ordnance needs and opted for the two 100lb incendiary bombs and four 100lb high explosive bombs.

Lukas and his crew departed Guadalcanal on the morning of 9 January and flew to Munda, on New Georgia Island, where they ate lunch. The trio continued to Bougainville in the late afternoon and landed at Torokina fighter strip.[2] At 10:00 PM the P-70 departed Torokina and proceeded on a course toward New Ireland, which was situated east of New Britain and Rabaul. Flying at an altitude of 500 feet, they spotted an American PT boat and, thirty minutes later, a dingy with a man rowing. Since the dingy was southeast of New Ireland Lukas was not sure if the man was an American or Japanese pilot who had been downed earlier. Operating under radio silence, the crew could not report the sighting until the next day after completing the mission. As the P-70 approached the coast of New Ireland one of the crew members spotted a Japanese aircraft, either a Mitsubishi F1M Pete or a Nakajima E8N Dave, at 1,000 feet.[3] Lukas quickly banked his aircraft and managed to slip away from the Japanese plane, believing that he was not spotted.

The crew made landfall on New Ireland at Cape St. George and flew north along the western coast of the island. Rain and poor visibility took its toll as they attempted to ascertain their exact location. Thinking he was directly east of Cape Tavui, New Britain, Lukas banked his aircraft and flew westward. However, upon reaching land he was not sure of his exact location and flew west along the coast for five miles before getting his bearings. The lack of any mountains to the south convinced Lukas that his arrival had been Cape Gazelle and not Cape Tavui. Lukas flew to the north and made landfall at Bolemakas Bay. The weather had cleared, and he set his course again for Cape Tavui. However, the weather worsened as he approached the northern coast of New Britain. Lukas identified Watom Island with difficulty and knew he was on course. The crew cruised offshore between Talili Bay and Tavui Point to study the weather situation over the target area. After careful contemplation of the situation Lukas opened his bomb bay doors and approached Rabaul from the direction of Vala Vala. Climbing over the ridge ringing Rabaul, Lukas dropped down toward the harbor as two searchlights snapped on and began swinging wildly in an attempt to locate the aircraft that had just passed their positions. The P-70 continued on a northerly course, hugging the terrain until reaching the harbor.

Lukas gunned the Douglas P-70 Nighthawk's throttles and initiated his bombing run at 300mph. The shadowy palms quickly slipped past the night fighter and gave way to the open water of Simpson Harbor. Despite the poor visibility due to the low ceiling and blackout conditions on the ground, the target could be seen lying at the distant end of the harbor. The pilot dropped the nose of the P-70 and skimmed above the quiet water as he lined up his final approach to the target. The absence of searchlights and anti-aircraft fire assured the crew that their mission was a complete surprise to the Japanese based in and around the town. Lukas could see the row of naval docks nearing his plane. According to intelligence photographs, a fuel dump and the Australia House lay just beyond the docks. Although the fuel dump was an obvious target for the night fighter crew, the Australia House offered a more tempting objective. The former elegant hotel served as a Japanese headquarters and/or officers' quarters according to intelligence sources. The P-70 buzzed across the naval dock at less than 100 feet altitude as the bombs were released. Delays on the bomb fuses allowed the P-70 to escape before the ordnance exploded over the targets. DeForrest yelled that the incendiaries had burst and produced a bright blue flame as they and the high explosives enveloped the dock, fuel dump, and Australia House, which he claimed actually rose from its foundation and collapsed. The fuel dump burst into flames as the incendiaries smothered their target. In addition, the pilot believed that one of the bombs might have damaged a small naval ship at the dock. Lukas immediately urged his P-70 into a climb and turned to cross the ridge located northwest of the town as the bombs exploded below him. Two Japanese searchlights situated on the ridge illuminated the dark night in a vain attempt to locate the attacker or attackers. The gunner aboard the P-70 swiveled his .50 caliber machine gun and silenced one of the lights as the plane roared across the ridge and dropped to the other side. Two additional searchlights swept the darkness above the silent water in search of possible naval craft and briefly illuminated the P-70 before losing the aircraft due to its speed.

Lukas flew east and crossed Duke of York Island on a course for New Ireland. Reaching New Ireland he turned south, and within minutes one of the crew members spotted a Japanese fighter above and dead astern of the P-70. Lukas decreased his altitude and skimmed the ocean in an attempt to lose the fighter. Evasive action failed to shake the Japanese fighter, so he headed inland

An American bombing attack over the Japanese base at Rabaul, on New Britain. The white spots in the harbor are bombs splashing into the water. When Captain Al Lukas made the first American solo night fighter raid on Rabaul, he approached from the right side of the photograph and utilized the long pier in the left-center as an aiming point for his attack. *Courtesy of AFHRA.*

and maneuvered his large aircraft into one of the valleys on the island. This action, placing the darkly-painted P-70 into the dark valley, apparently confused the Japanese pilot, allowing Lukas to successfully slip away.

As Lukas, Blankenship, and DeForrest approached Cape St. George they decided to conduct an unscheduled harassment of the Japanese. Allied forces knew that the Japanese operated an anti-aircraft and radar installation approximately two miles inland from Cape St. George.[4] Since the weather in the area was clear, Lukas opted to locate and strafe the Japanese installation. After making a dry run at an altitude of 200 feet the trio spotted the tents, anti-aircraft gun positions, and radar antenna in a clearing. Lukas guided the P-70 to the south and returned for his strafing run. As he approached the installation, he fired a three-second burst from the six .50 caliber machine guns in the nose of the P-70 at the tents and gun positions. Returning again, Lukas fired a burst of machine guns into the radar shed while Blankenship, on the dorsal twin .30 caliber guns, and DeForrest on the ventral .50 caliber gun peppered the tents. Lukas returned for a fourth pass over the installation and noticed a white light flashing approximately two miles off shore. Suspecting that the light could be from a Japanese surface vessel approaching Cape St. George in response to an emergency call from the installation, Lukas returned to his original flight path and headed south toward Bougainville.

Lukas eventually climbed to 9,000 feet to avoid poor weather and safely landed at Torokina at 2:00 AM, four hours after commencing the raid from the same location. He attributed the lack of anti-aircraft fire, despite at least twenty Japanese sites along the ridges surrounding Rabaul, to the complete surprise of the low altitude night mission. Lukas received the nickname "Super Luke" from the raid. After landing and reporting the success of his mission, the unit intelligence officer, Captain Warren "Skip" Slater, told him, "That was super, Luke." Lukas carried the nickname for the rest of his life.[5]

Sadly, another night fighter crew attempted the same mission on 18 January despite warnings to abandon the idea. Based upon his RAF experience against the Germans, Lukas realized that the Japanese would be alert for a repeat of the caper. Apparently, the Japanese were ready for the possibility of another low altitude night raid and the crew attempting to duplicate the mission never returned from the operation. The P-70 on this Rabaul mission was officially listed as Missing in Action, and the Air Force has never been able to locate the aircraft or crew. The missing crew members are Second Lieutenant Talbot Kelley (pilot), Flight Officer James Palmer (R/O), and Staff Sergeant Kenneth Nicholson (gunner). After losing this crew, the 13th Air Force night fighters did not attempt any additional solo night intruder attacks on Rabaul.

The 419th NFS lost another P-70 and crew the day before Lukas flew to Rabaul. A former Detachment B pilot, Captain Harry Wayne, perished along with Sergeant Edward Tomlinson and Sergeant Richard Bennett in a P-70 day landing on Guadalcanal on 8 January. This was the third loss of a P-70 since 22 November when Captain John McCloskey crashed during a training mission. Two days after the loss of Captain Wayne and the others, and the day after the night intruder raid on Rabaul, the 419th NFS lost a P-38 during a takeoff attempt. First Lieutenant Dick Stewart teamed with First Lieutenant Francis Parker for a P-38 mission on 10 January from Guadalcanal. Stewart and Parker took off as a pair down the airstrip. As Parker's P-38 sped down the airstrip one of the aircraft's engines failed. Stewart noted that he looked over in time to see Parker's plane crash. Stewart managed to slide his plane to the left to avoid the wreckage. Although the P-38 was totaled Parker survived the accident.[6]

The 419th NFS conducted four types of missions as they awaited the arrival of their Northrop P-61 Black Widow fighters. The raid on Rabaul by Lukas and his crew can be referred to as a night intruder mission. This combat operation category involves a night attack against a specific target with the purpose of damaging enemy facilities and escaping before enemy defenses can pinpoint your location. The second type of mission was night harassment. Normally, the night crews do not have specific targets, but are doing their best to inflict any damage, as well as working to destroy the morale of the enemy forces on the ground through sleep deprivation and concern for their safety. This is the same type of mission flown by "Washing Machine Charlie" and "Louis the Louse" over the Allied positions on Guadalcanal. Former Japanese fighter pilot Masatake Okumiya served at the Buin airfield on Bougainville between early July and late September 1943. Okumiya, in his classic 1956 book *Zero!*, co-written with Jiro Horikoshi and Martin Caidin, discussed the morale destroying impact of American night harassing missions:

"Maybe if we are lucky, it will be a restful evening. Before long we know differently…Perhaps there is only one [American] airplane up there, circling the field, dropping a bomb every now and then, but he keeps us awake for hours…"[7]

419th NFS Intelligence Officer Captain Warren Slater (without flying cap) briefs four of the unit's pilots. Standing left to right are First Lieutenants Marvin Walker and Don Dessert. Kneeling left to right are an unknown pilot and First Lieutenant Dick Stewart. The squadron headquarters remained on Guadalcanal and a detachment operated from Bougainville. Air and ground crews departed Guadalcanal to serve brief tours with the detachment on Bougainville. 419th NFS aircraft also pulled night alert duty at other locations, such as Munda and Green Island. *Courtesy of Leslee Lukas Mays.*

Night fighter defense was the third type of mission flown by the pilots of the 419th NFS. Primarily, these were the P-38 search light operations flown over Guadalcanal, New Georgia, Bougainville, and later Green Island by the night fighters in defense against Japanese night intruder and night harassing missions. The Navy's VF(N)-75 and Marine's VMF(N)-531 flew night defense missions over the many vessels supporting military operations in the Solomon Islands, while the 419th NFS of the 13th Air Force provided night protection over Allied land bases via P-38s working in cooperation with ground based searchlights.

The fourth type of mission was the day harassment of Japanese forces. The 419th NFS flew many of these missions against the Japanese on Bougainville from December 1943 through the end of May 1944 when they finally transitioned to the P-61 Black Widow. During this period the pilots pulled night alerts at Bougainville's Torokina and Piva airfields awaiting a possible attempt by the Japanese to sneak a raider into the area. The next morning, the pilots would take off and spend 60-90 minutes flying over Bougainville's Japanese held areas looking for targets of opportunity, which ranged from Japanese patrols to facilities. After strafing any targets they spotted the pilots flew their P-38 Lightnings back to their basing areas, which could be Guadalcanal or Munda, on New Georgia. By February 1944 the 419th NFS maintained a detachment on Bougainville, permitting the P-38s to fly daylight harassing operations and return directly to Torokina. Naturally, a pilot could

Left to right: Flight Officer Francis Meyer (radar observer), First Lieutenant John McGhee (pilot), First Lieutenant Ralph LeVitt (pilot), Flight Officer George Black (radar observer), and First Lieutenant Bob Frenzel (pilot) of the 419th NFS. When the unit received its P-61 Black Widows Meyers and McGhee flew as a pilot/radar observer team, while Black and Frenzel flew as another team. Levitt flew with Flight Officer Freddy Kahn. *Courtesy of Don Dessert.*

fly a combination of the four mission categories during the same sortie.

Second Lieutenants Don Dessert and Dorval Brown made a typical 90 minute day harassing flight over Bougainville on 10 January, the same day Lukas flew the night intruder mission to Rabaul. Dessert and Brown took off from Torokina airfield at 6:30 AM and proceeded directly over the mountains that run through the central part of the island. The pair banked and began a patrol along the east coast of the island. Dessert flew 50 yards to the right and 100 yards behind Brown as the pair searched for Japanese targets of opportunity. The pilots proceeded along the coast at 270 miles per hour at altitudes that ranged from tree top level to 300 feet while strafing any grass huts they spotted. Near Arawa Plantation they strafed three wooden vehicular bridges which were well camouflaged with natural foliage. At Kieta Peninsula the pilots fired upon a white house as Dessert spotted six tracer rounds arching up and past his right wing from an unknown position on the ground. Several miles south of Cape Friendship the duo spotted two Japanese soldiers bearing rifles and carrying large packs fleeing into the woods. Dessert and Brown made a 270 degree turn and thoroughly strafed the woods into which the Japanese soldiers had fled. The men were probably the rear element of a larger party of Japanese troops. The results of the attack could not be ascertained by the pilots, who climbed to 10,000 feet and returned to Guadalcanal, where they landed at Fighter Strip #2 at 9:00 AM.[8]

Second Lieutenants Dessert and Brown flew another harassing mission three days later on 13 January during which Brown believed Dessert had been shot down and lost on the eastern end of Bougainville. The pair departed Torokina in P-38s at 7:45 AM and proceeded toward the southeast at 2,000 feet altitude. At Toko Mission, the pilots initiated a strafing run along the beaches in search of targets of opportunity. Brown flew along the shoreline while Dessert maintained a parallel course approximately ½ mile inland. As Dessert cruised at 250 miles per hour he fired occasional bursts from his machine guns at grass huts passing beneath his P-38.

First Lieutenant Don Dessert. The 419th NFS flew almost daily P-38 Lightning daylight harassing missions over Bougainville between January and May 1944 as the unit awaited its long promised P-61 Black Widow night fighters. Dessert flew many of these operations. On 13 January Dessert surprised and strafed approximately 20 Japanese soldiers working in a garden on Bougainville. *Courtesy of Don Dessert.*

The 868th Bomb Squadron, "Snoopers" flew Consolidated SB-24 Liberator bombers mounting radar to locate and attack Japanese vessels at night. If they did not find any Japanese vessels, the "Snoopers" would fly over known Japanese bases and drop their bomb load before returning. Many of the 419th NFS Radar Observers flew with the "Snoopers" while the unit awaited its P-61 Black Widows. *Courtesy of AFHRA.*

The pilots rounded Moila Point and proceeded north to the Japanese held Kahili Airfield. Dessert increased his speed to 320 miles per hour as he approached Kangu Hill, located north of Kahili. A group of approximately twenty Japanese soldiers wearing white sunhats, no shirts, and long pants were working in a garden near Kangu Hill. The Japanese soldiers were intently working their garden and did not notice Dessert as he approached them. Dessert fired several bursts from his machine guns and 20mm cannon, catching them totally by surprise. The pilot remembered seeing the soldiers running wildly before falling as he whizzed by them. Later he estimated that he dropped ten of the Japanese soldiers in that single pass. Dessert banked his aircraft and headed toward Kahili. Noticing tracers skimming past his left wing, Dessert dropped his altitude and flew south for approximately 15 miles, at which point he climbed to 5,000 feet and returned to Guadalcanal at 10:15 AM.

Brown, after passing Kahili, noticed that Dessert was not flying off his left wing. He did not realize that Dessert had dropped his altitude and engaged the Japanese soldiers at Kangu Hill. Brown increased his altitude and initiated a search for his wingman, thinking that he might have been shot down by anti-aircraft fire along their flight path. He spotted a white parachute in some trees along the shore and thought the worst had happened to Dessert—that he had been shot down and forced to parachute into Japanese held territory. The pilot radioed Torokina to report that he had a man down. Fighter control at Torokina replied that the parachute had been noted the previous day, so it could not be Dessert. Brown concluded that Dessert must have been distracted and then returned to Guadalcanal. He increased his altitude to 15,000 feet and flew to Munda, where he landed at 10:00 AM, 15 minutes prior to Dessert at Guadalcanal.[9]

Two days later Dessert flew a solo harassing mission to Luluai Point on Bougainville. Departing Torokina

Left to Right: Flight Officers John Radzieta and Roger Egan. Radzieta and Egan flew missions with the "Snooper" SB-24s while the 419th NFS awaited its P-61 Black Widows. Radzieta remembered, "...We never came home with bombs aboard. If we found no shipping, we dropped our ordnance on Japanese installations on New Ireland, giving them a 'piss call.' It was the American answer to 'Washing Machine Charlie.'" *Courtesy of Don Dessert.*

at 7:45 AM, Dessert proceeded at 4,000 feet to Tarara village, where he dropped his auxiliary gas tank and strafed with unobserved results. He made a second strafing run and then proceeded south along the coast. Dessert noticed human tracks in the beach sand which appeared as columns of four, indicating the passage of a large body of Japanese soldiers. The pilot proceeded along his flight path and spotted a wooden bridge. He turned and fired several bursts of his machine guns and cannon into the bridge. Dessert continued his mission and occasionally strafed known Japanese targets, including Kieta Airstrip. From Bougainville Dessert flew south to his home roost on Guadalcanal, landing at 10:15 AM.[10]

With military operations being flown in P-38s the R/Os of the 419th NFS lacked a mission, since the fighters were single seaters and did not carry onboard radar. The solution was to loan the R/Os to the 868th Bomber Squadron (BS), which flew SB-24s out of Munda. SB-24s were Consolidated B-24 Liberator bombers equipped with airborne radar. Their primary mission included utilizing radar to search and locate Japanese naval vessels, which they would then target with their bomb loads. Ten of the modified bombers, nicknamed Snoopers, arrived in the Southern Pacific area in August 1943 and were designated as the 394th Bombardment Squadron of the 5th Bomber Group based at Guadalcanal. The SB-24s proved successful throughout the fall of 1943, resulting in the establishment of two independent squadrons: one for the 5th Air Force and the other for the 13th Air Force. The SB-24s of the 13th Air Force were redesignated as the 868th Bombardment Squadron in January 1944 and operated from the airfield at Munda, where the 419th NFS stood night alerts.

Each SB-24 carried an SRC-717-B Air-to-Surface Vessel (ASV) search and navigation radar, an AN/APQ-5 LABS bombing radar, a SCR-729 IFF (Identification, Friend or Foe) device, and an AN/ARN-1 radio altimeter. The latter proved to be a useful safety tool during the long flights over water at night and assisted with bombing accuracy. Crews mounted the antenna for the SRC-717-B in the bomber's ball turret underneath the aircraft. The SB-24s patrolled at night in search of Japanese vessels transporting men and supplies between their bases in the area. The SRC-717 radar system searched for vessels moving any distance away from the shore. Upon detection of a target utilizing the radar, the crew would turn to the SCR-729 to see if the vessel was emitting an IFF signal, indicating it was an Allied vessel. The R/O would direct the pilot for a bomb run over any unfriendly vessel. If the crew had not located an enemy vessel during the night they would dump their ordnance over a known Japanese base before landing. Although the number of vessels sunk by SB-24 crews was exaggerated during the war due to the fact that they could not actually see the sinking of a vessel at night and had to rely on secondary evidence, such as the disappearance of a radar "blip" following an attack, they were very successful in locating and striking enemy shipping.

Initially, seven R/Os from the 419th NFS were designated to serve with the 868th BS. They were, in turn, replaced by a second group of night fighter R/Os. The 419th NFS R/Os found the missions useful and a good application of their skills while their squadron awaited the arrival of the P-61 Black Widows. Flight Officer John Radzieta, one of the first seven 419th NFS R/Os to fly with the Snoopers, noted that his job was to relieve the regular R/O during the very long night missions in the vicinity of Rabaul:

"All five of my missions with the Snoopers were routine. We relieved the radar operator and navigator for spells while looking for Japanese shipping to New Britain—mainly Rabaul. It was good duty and good training for tuning the radar. When we spotted a target we aroused the members of the regular crew, who were school trained on the particular systems mounted in the Snooper aircraft. They got the glory in making the bomb run. We dropped bombs on shipping, and it often took two or three passes before the blip on the radar scope disappeared. On three of my missions we found no shipping. It was a ten hour patrol in the Bismarck Sea and we never came home with bombs aboard. If we found no shipping, we dropped our ordnance on Japanese installations on New Ireland, giving them a 'piss call.' It was the American answer to 'Washing Machine Charlie.'"[11]

Flight Officer Pete Fulghum also helped initiate a "piss call" on a Japanese camp while flying as an R/O on a Snooper. Fulghum remembered participating on a Snooper mission to Rabaul during a period of poor weather. The SB-24 pilot opted to turn around due to the worsening weather conditions. While looking for an alternative location to drop his bomb load the pilot spotted lights on a small island:

"He dropped us down to just over the ocean and we flew in really low toward the lights. He pulled up just over the treetops and we came in over a Japanese camp. I had my head out and saw Japanese soldiers running everywhere, and the bombs were falling right in the middle of the camp. No telling how many we got that night!"

Flight Officer Roger Egan also participated in the 868th BS Snooper missions with the first group of seven 419th NFS R/Os:

"It was my turn on the scope. We picked up this blip at thirty miles. I awakened the regular crew radar operator. We turned toward the target. The blip grew stronger. We identified it as a large ship. We started a bomb run. Ack ack opened as we approached the target. It looked like fireworks on Independence Day. I was looking out the waist window. Some of the flak was popping so close that I could have reached out and lit a cigarette. Instead of getting a 'Bombs Away' from the bombardier we got an 'Abort.' We broke off the run 'Whew' I breathed, 'I'm glad that's over.' The bombardier stated that the bomb bay doors did not open, disabling the drop. The pilot directed the flight engineer through some troubleshooting procedure. This failed. He announced that the doors would be hand cranked open. But first he called for me (the spare radar man) to come forth to the cockpit. I did. He motioned me to come closer. As I did he removed his oxygen mask and shouted into my ear. 'During this next run, your duty station is between the co-pilot and me. If either of us is hit by flak it will be your duty to drag the hit person out of the seat so that he cannot jam the controls.' I got to thinking, 'Here I am within a foot of each pilot. If one is hit, I probably will also be hit.' I tried to push back the bulkhead to get more distance but it would not move. Talk about sweating and aging. The third run was successful. We left a burning ship. Upon return to Munda our plane was grounded for sheet metal work."[12]

Flight Officer Jim Allen also flew snooper missions with the first group of 419th R/Os seconded to the 868th BS. During one flight Allen flew as the primary R/O aboard the SB-24:

"Every night two or three SB-24s were sent up to cover the waters around Rabaul and to bomb any Japanese ships bringing supplies to the base. I had a couple training missions and then was on my own as the sole R/O for the next one. We were cruising along when I picked up this radar target. I checked my charts and found no islands or anything else that could have been there to provide a radar return on the scope. I called the target in and gave the pilot a vector to intercept it. We had a full moon, so the bombardier should have had a visual sighting in a few miles but no one up front could

Flight Officer Jim Allen (second from right) was another 419th NFS Radar Observer who flew missions with the "Snooper" SB-24s. During one operation over a Japanese base, Allen leaned out the opening for the waist gunner in order to see what was happening on the ground during the attack. He nearly tumbled from the aircraft as the pilot made a violent maneuver to avoid a Japanese searchlight. "After a mission, you are entitled to a couple shots of combat whiskey, but after being up all night and not yet having breakfast, no one took it. This time I took it and got another from a buddy." *Courtesy of Leslee Lukas Mays.*

see the target. Then it dawned on me that I had not taken into account that it could have been an airborne target rather than a ship on the sea. Our radar sets did not provide information on altitude, just direction. Everyone up front was yapping and I couldn't break in to tell them my possible error. In just a few seconds someone on the intercom said, 'It's an airplane and he is blinking his lights.' The top gunner said, 'Blinking his lights, hell. He's shooting at us.' We made some turns and saw them make some turns on the radar. Since we were going in opposite directions we separated very quickly. I watched them on radar, and they made a couple more turns and continued south. I think there were three Japanese aircraft in that formation. I got some kidding over this, but it was exciting."[13]

Allen experienced one additional memorable mission with the Snoopers:

"In another incident, we had not found any shipping so we went over to New Ireland to bomb the airfield. This was done by hitting an initial point, taking a pre-determined heading and airspeed, and dropping the bombs in so many minutes. It was my job to get the aircraft to the IP ["Initial Point" from which the pilot begins the bomb run]. After doing this I went back to the right waist window and leaned out to see if I could see the bombs hit. This window is pretty large. By this time we were getting quite a bit of close ground fire. I don't know if it was luck or not, but a searchlight came on and hit us square. At

Left to right: Flight Officer John Radzieta, First Lieutenant Aaron Jacobs, and Flight Officer Sandy Bauman on leave in New Zealand. Air crews received brief leave periods in New Zealand or Australia as a form of combat stress relief. Popular events while on leave included eating anything fresh and not canned; checking out the local ladies; and stocking up on liquor to take back as trade goods. Pilots rarely forgot to take care of their aircraft crew chiefs while on leave and returned with gifts of liquor and other items. *Courtesy of John Radzieta.*

this point I was not wearing a parachute. When the searchlight hit me in the face I thought I might have been shot. The pilot immediately made a right, hard sloppy turn to get out of the light and I started out of the window. I threw my right arm up and managed to catch the edge of the window, and had my knees on the lower part, but was swinging in the slipstream, looking straight down. I knew I could not hang on and had the fleeting feeling I was already falling. It all happened very fast. Luckily, the pilot made a hard left turn just before I lost my grasp and I landed in a heap on the far floor. After a mission you are entitled to a couple shots of combat whiskey, but after being up all night and not yet having breakfast no one took it. This time I took it and got another from a buddy. I still remember it was Old Overholt, rye whiskey. Also, for a couple of years after the war when someone took a flash picture with those bulbs and large reflectors I would flinch."[14]

Each 419th NFS R/O in the initial group flew 10 combat missions with the Snoopers. Everyone noted this initial group of R/Os performed well, and the 868th BS appreciated the assistance. A second group of 419th NFS R/Os also served with the Snoopers as their pilots awaited the long promised P-61 Black Widows, which would return everyone to their flight teams. The completion of 10 combat missions qualified each R/O for an Air Medal and a 10-day Rest and Recreation (R&R) in New Zealand. The 419th NFS pilots flying the P-38 and occasional P-70 combat missions over and around Bougainville qualified for an Air Medal and R&R as well. On average, each pilot and R/O received a 10-day R&R approximately every 3 months to help them with the stress of flying combat missions. This often meant that the air crews actually flew more than 10 combat missions before receiving the R&R since only a small percentage of them could be on leave at one time and their return could be delayed due to combat requirements for the transport aircraft.

The ground crews of the 419th NFS demonstrated a keen appreciation of the dangers faced by their air crews. First Lieutenant Robert "Bob" O'Brien, the squadron ordnance officer, compared the dangers he faced on the ground with those of the squadron's pilots:

"Sure we got bombed a few times, but every day they flew out over the ocean on their missions and some did not come back. I got a small taste of their life when I flew as a passenger in a P-70 from Guadalcanal to Bougainville. We were the target of antiaircraft fire from an island on the southern tip of Bougainville and had to make a quick dive to about 39 feet off the water, and when we landed at Bougainville we landed short in tree stumps and blew out the tires. The pilots can go through this every day and that would be too much for me."[15]

Rumors and folk myths abounded on Guadalcanal after the arrival of the 419th NFS. Sergeant Jesse Gardner, an Operations clerk with the 419th NFS, recorded several in his diary. At the time he believed them, since there was not a reason to question their authenticity. However, today they can be seen as quite amusing. One rumor stated that Japanese soldiers left behind in the area were so hungry that they would sneak through the American lines and stand in chow lines with the American personnel for a meal. An interesting myth claimed that an American latrine on a particular ridge was built over the burial site of General Hideki Tojo's son. (Hideki Tojo was the Prime Minster of Japan from 1941 to 1944). Tojo's son supposedly died on the 13th of an unknown month in 1942, and the Japanese launched air raids on the 13th of every month in memory of their Prime Minister's son. Gardner, expecting this to be a true event, noted in his diary that the Japanese seemed to have missed their monthly raid on 13 November. Another story, probably told and re-told to frighten newcomers to the island, related how an American sitting on a latrine was knifed by a Japanese soldier sitting on the latrine next to him.[16] While the myths obviously concerned newcomers to Guadalcanal during the war, today they certainly offer us an amusing glimpse of wartime culture.

Beer was perhaps one of the most important icons of American culture for the servicemen overseas. Beer represented something difficult to acquire, as well as memories of their peacetime lives back home. When available, beer was provided to the men in monthly rations and was so important for morale that the number of bottles authorized in each issue was dutifully recorded in the monthly squadron historical report. These reports meticulously noted the transfer of personnel, status of aircraft, important combat missions, and the amount of beer authorized for issue to each squadron member. Beer could be worth more than cash. Flight Officer James "Jim" Allen remembered, "I got a nice Japanese Officer's sword for two beers. Beer was much better than money out there."[17]

Arriving on Guadalcanal in November 1943, the 419th NFS personnel encountered the phenomena known as the "beer ration," and being resourceful, they began thinking of how they could increase the amount of beer available to them. This led to what many 419th NFS veterans have referred to as "The Beer Caper." Beer arrived on Guadalcanal via cargo ships, which transferred the bubbly commodity via cargo nets attached to cranes. Many times, these nets broke or the military personnel did not properly balance the loads inside the nets, which resulted in crates tumbling out and falling into the water. The 419th NFS surmised that there had to be bottles of beer lying beneath the water just waiting to be claimed. Bill Mackie designed and constructed a diving bell from aircraft wing tanks. The bell could be fitted over the head and reached down to the hips of an individual with a small frame. Compressed air would be pumped into the top of the cylinder doubling as the helmet. As long as the bell was kept in the vertical position it produced a pocket of air for the man wearing it. A hole was cut in the front and the metal replaced with Plexiglas to serve as a portal for the diver.

The bell was so small that only two men could wear it: First Lieutenants Dick Stewart and Robert "Bob" Winslow. Stewart, the youngest pilot in the unit, wore the bell on a beer salvaging mission. The bell was a difficult device to wear, since he had to maintain the pocket of air and this prevented him from stooping over. He did manage to carefully pick up bottles of beer and food from the bottom of the harbor area and stuff them into mattress covers for the others to haul up from the water. To increase the amount of beer available for scavenging the officers bribed the military stevedores with brandy from Australia to "accidentally" drop a few cases of beer as it was being transferred to shore. A Navy officer in charge of the dock area discovered what the 419th NFS was attempting and forced them to stay away.

A group of USO entertainers asked to visit the 419th NFS' new officers' club following the failure to remove the beer from the harbor. The resourceful night fighters offered to trade brandy acquired in Australia for beer. John Radzieta picks up the story:

"Four of the R/Os who spearheaded the modifications to our club volunteered to donate a bottle of brandy each and seek enough beer to have a successful party for the USO group. They found a 'beer broker' and made a deal. The truck drove up into the officers' area with more cases of beer than they realized were available. They did not want to renege on the deal so they ended up biting off more than they could chew. The R/Os scrounged enough brandy and accepted the full load of beer. The party was a great success, but they did not have a place to store the remaining beer."

Flight Officer Sanford "Sandy" Bauman commented that it looked like the unit had developed a beer distributorship.

First Lieutenant William "Bill" Mackie, one of the squadron's former Royal Air Force (RAF) pilots, found a photographic dark room and placed the unit beer supply in this facility. Major Emerson Barker, the unit commander, discovered the excess beer and ordered it returned. The Military Police caught the men in the process. Don Dessert noted, "Our mistake was returning the beer in a 419th truck." The local Inspector General's office became involved in an investigation, and the remaining beer supplies had to be returned. The military sought someone to blame for the caper, and Mackie was the highest 419th NFS ranking officer on Guadalcanal at the time who was involved in the beer acquisition project. Mackie was court martialed and discharged from the service upon completion of his combat tour. Every squadron member viewed the court martial as an injustice. In their eyes, Mackie had been made a scapegoat for the events. After his discharge Mackie enlisted in the Army Air Force to help clear his name from the incident and remained in the service until retirement.[18]

Another view of 419th NFS' B Flight. Left to right (in front): unknown, First Lieutenant Bob LaCroix, First Lieutenant Al Lukas, Flight Officer Sandy Bauman, First Lieutenant Bob Winslow, and Flight Officer Lane Thompson. Left to right (in the rear): First Lieutenant John Blankenship (standing in shirt), First Lieutenant Don Dessert (leaning on bumper), Second Lieutenant Phil Rand (in cap), Flight Officer Fred James, Flight Officer Bill Riley, Flight Officer Malcolm Holloway, First Lieutenant Herman Manke (in shirt), and First Lieutenant Jim Schroth (in shirt). *Courtesy of Leslee Lukas Mays.*

First Lieutenant Al Lukas (leaning on the left front bumper of the vehicle with his hands on his legs) and the air crews he led as part of the 419th NFS' B Flight. First Lieutenant Don Dessert is in shorts and leaning on the right side of the vehicle bumper. Upon arrival on Guadalcanal the squadron consisted of co-located flights. Later, it would form separate detachments as the 13th Air Force spread the unit to cover various islands in the advance to the Philippines. *Courtesy of Leslee Lukas Mays.*

Flight Officer Frank Hartwick died in a tragic accident on February 25, 1944, on Munda. Hartwick took off from Munda in a P-38 to intercept a Japanese raider in cooperation with the local searchlight crews. The pilot was not able to locate the bogey so he returned to land. It is speculated that he confused the beach for the coral landing strip in the dark. Flight Officer Jim Allen, a 419th NFS radar observer flying with the "Snoopers" out of Munda, attended Hartwick's burial and later noted, "I don't think I have ever been so sad and depressed." *Courtesy of Don Dessert.*

Movies were another important pastime of the night fighters. The 419th NFS established an outdoor movie theater at all of its permanent bases between Guadalcanal and the Philippines and showed old as well as fairly recent films sent from the United States. "It did not matter what movie we saw because they all reminded us of home and the ones we left behind. It didn't matter if it rained, we just sat on our log and watched the movie to the end, then went back to our tent and dried off and went to bed," explained First Lieutenant Robert "Bob" O'Brien. At times illicit "X" rated films made the rounds. O'Brien remembered one poor quality, grainy "X" rated film that crisscrossed the American movie theaters on Guadalcanal:

"It was quite tame by modern standards, but was nonetheless against regulations. That movie had two young women in it who came out to a pool and took off their robes to reveal they were wearing the equivalent of a bikini. They strolled around, then jumped in the pool and the movie ended. There were at least 12 movie houses on Guadalcanal. By movie houses I mean a large movie screen tied up among the coconut trees and rows of coconut trees for seats. Once we saw this movie with the women we were eager to see it again, or hopeful that more would show up, so we played a guessing game and tried a different movie each time, hoping that we picked the right one that was showing it. If the Military Police arrived the film would be pulled off the projector upon receipt of a signal and passed to a man who would vacate the premises. The next night the film would show up at another unit."[19]

First Lieutenant Stewart managed to avoid a catastrophe when returning to Guadalcanal following a mission over Bougainville. His P-38 suffered anti-aircraft weapons damage to the right wheel well, a lack of hydraulics, and the loss of the tire while flying over Bougainville. Ground control at Guadalcanal offered Stewart the option to belly land or bail out over the water. Stewart chose the latter option, since a belly landing would damage the air strip and force engineers to make repairs. In addition, he had always wanted to make a parachute jump and this was his opportunity. The key to bailing out of a P-38 involved reaching an altitude of at least 3,000 feet, opening the canopy, turning the ship over, and allowing gravity to

The Allied air strip at Munda, on New Georgia Island in the central Solomon islands. The 419th NFS provided night alert aircraft here while awaiting the arrival of its P-61 Black Widows. Flight Officer Frank Hartwick died in a P-38 accident attempting to land at Munda. *Courtesy of the U.S. Army.*

pull you out and clear the rear stabilizer. As Stewart made the first pass something told him not to jump. He turned the aircraft and made a second attempt. Again, something told him not to leave the aircraft. The pilot radioed that he would attempt a belly landing in lieu of a parachute jump over the water. He successfully bellied the plane and afterwards returned the parachute to the riggers. Standard procedures called for a chute to be replaced after an accident. As the riggers opened the equipment the silk simply fell apart in their hands. The excessive moisture on the islands had rotted the parachute silk. If Stewart had bailed out of his P-38 the parachute would not have deployed properly and he would have been killed in the 3,000 foot fall.[20]

On 12 February the 419th NFS lost its last P-70 Nighthawk during a crash landing. First Lieutenant Don Dessert shuttled the P-70 from Guadalcanal to Bougainville with Flight Officer Lane Thompson and a load of mail, food, and beer for the squadron's Detachment A on the island. As he approached the airstrip Dessert reached to flip the actuating handle for the landing gear. The handle came off, leaving him holding it in his hand. With no way to lower the landing gear Dessert was forced to belly land the P-70. The resulting slide pierced the steel plank airstrip and totaled the P-70. However, many veterans of the 419th NFS have boasted over the years that Dessert did not break a single egg or bottle of beer in the landing.[21]

Successful Allied raids on Rabaul and other air bases in the area demonstrated the Japanese had lost control of the skies as well as the aircraft replacement route through Truk, an island north of Rabaul. Masatake Okumiya, transferred to Rabaul as a staff officer with the Japanese military on January 20, 1944, lamented the sharp change in the war. He found Japanese air personnel apathetic over their plight at Rabaul and sought methods

First Lieutenant Dick Stewart pulled the P-38 night alert shift on Green Island, just north of Bougainville, on several occasions. A Navy logistics lieutenant on Green Island was well known for providing air crews with comfort items, such as hamburgers, milk, and coffee. Stewart deeply appreciated the tireless efforts of this young officer named Richard Nixon, who would later become President of the United States. President Nixon fondly remembered his days on Bougainville and Green Island when writing his memoirs later in life. Stewart later flew the election campaign aircraft for Nixon's running mate, Spiro T. Agnew. *Courtesy of Dick Stewart.*

to improve the situation. Following orders in late January to only engage American aircraft under favorable conditions, Okumiya increased the number of Japanese night intruders dispatched to raid American bases and vessels."

"To counter the effects of the new order, I reinstituted small night offensive raids...At no time did I believe that the night attacks would affect noticeably the enemy's northward drive, but the feeling of returning to the offensive would improve morale...We could not muster more than twenty planes to use in these raids and the results, of course, were neglible."[22]

Night fighters of Marine Corps VMF(N)-531, flying from Bougainville and later Green Island, bagged five of the twenty Japanese night intruders dispatched by Okumiya between 5 and 17 February. The 419th NFS of the 13th Air Force flew P-38 Lightnings without onboard radar systems during this period. Thus, they normally defended base areas in cooperation with ground-based searchlights and were not available to intercept Japanese night intruders over the ocean northwest of Bougainville.

After 17 February the Japanese realized they needed to withdraw most of their aircraft from the area in order to cut losses. The withdrawal necessitated a reduction in night intruders, which resulted in fewer opponents flying into the local hunting grounds of the Army, Marine Corps, and Navy night fighters.

The 419th NFS lost a pilot in an accident prior to the end of February. Flight Officer Frank "Eager Frankie" Hartwick crashed during a landing attempt on 25 February. Hartwick was the only pilot who arrived with the original group of the 419th NFS in November 1943 holding the rank of Flight Officer, rather than being a commissioned officer. He earned the nickname "Eager

Frankie" due to his energy and positive attitude, and was a popular and well liked pilot. Hartwick took off from Munda in a P-38 to intercept one of Okumiya's raiders in cooperation with the local searchlight crews. The pilot was not able to locate the bogey so he returned to land at Munda. It is speculated that he confused the beach for the coral landing strip in the dark. Flight Officer Jim Allen, a 419th NFS R/O flying SB-24 snooper missions from Munda, attended his squadron mate's burial:

"The burial was something I'll never forget. The weather was misty with a very light drizzle, cool and cloudy. The army had fought a battle through the area where he was buried. The few trees were stark with their leaves and branches shot off. The ground was muddy. They brought out Frankie in a small Army truck, and as we were taking him out of the truck one of the Army sergeants said, 'He's lucky – he's got a coffin.' We carried him over to the grave and a young Chaplain said the usual words that I'm sure he had said for many more Army dead. I don't think I have ever been so sad and depressed. Here we were burying this young, likable kid, and his family and friends were halfway around the world. At that time I guess we were his family. It just seemed very sad and unfair."[23]

The Navy's VF(N)-75 Corsairs departed from the Solomon Islands before March 1944 due to the Japanese reduction of fighter assets, while VMF(N)-531's Venturas continued to fly from Bougainville in order to protect the Navy vessels in the area of the that island and Green Island. During the same period, the 419th NFS commenced night alerts on Green Island. A P-38 pilot would take his plane to Green Island each afternoon and pull alert all night on the airfield. If a bogey approached, the P-38 provided coverage over the island while the Venturas patrolled over the sea and watched the American vessels. First Lieutenant Dick Stewart of the 419th NFS made numerous strip alerts on Green Island and met a Navy logistics lieutenant who befriended all of the pilots providing protection over the island. This lieutenant made sure the tired pilot pulling the night alert had comfort items, such as hamburgers, milk, and coffee. Stewart mentioned how he deeply appreciated the tireless efforts of this young logistics officer—a man who would later become President of the United States.

President Richard Nixon fondly remembered his days on Bougainville and Green Island when writing his memoirs. Originally assigned to duty on the more peaceful island of New Caledonia, he managed to land an assignment to a forward area and arrived on Bougainville in January 1944. Within the book Nixon provided insight into his support of the combat pilots like Dick Stewart staging through Bougainville and later Green Island:

"Many fighter and bomber pilots came through Bougainville on their way to battle missions, and I felt that they deserved the best we could possibly give them. I used my SCAT [South Pacific Combat Air Transport Command] resources to get small supplies of chopped meat and beer. Everyone in the unit had a nickname, and I was known as Nick Nixon. Whenever I received a fresh shipment, I opened 'Nick's Hamburger Stand' and served a free hamburger and a bottle of Australian beer to flight crews who probably had not tasted anything to remind them of home in many weeks."[24]

Years later Stewart, a commercial pilot, flew an aircraft leased by Richard Nixon's 1968 presidential campaign. Stewart's job was to fly vice presidential candidate Spiro Agnew around the country. After Nixon's election victory the newly elected president invited various individuals who assisted the campaign to a reception at the White House. When shaking hands with President Nixon, Stewart reminded the President that he had been one of the pilots who benefited from "Nick's Hamburger Stand" on Green Island. President Nixon gave an immediate hearty laugh that was captured by a White House photographer. An autographed copy was soon on its way to Stewart's home as a token of appreciation from the President.[25]

The 419th NFS' only kill during this period was scored by Dick Stewart, although it was never officially confirmed and thereby credited to the pilot and the unit. Stewart had been pulling strip alert on Munda and was returning to Guadalcanal. At approximately 9:00 AM he spotted the unmistakable silhouette of a distant Japanese Mitsubishi G4M Betty bomber heading northwest toward its home base on New Britain. Stewart turned his P-38 and initiated a pursuit of the Betty. The Japanese pilot saw the approaching P-38 and headed for a large bank of clouds in which he could hide. Stewart realized that the Japanese bomber would reach the relative safety of the dense cloud bank before he could get close enough for an effective attack. He knew he needed to fire upon the escaping bomber even though considerable distance remained between the two aircraft. Firing upon the

First Lieutenant Dick Stewart made an unclaimed kill of a Japanese Mitsubishi G4M Betty bomber while flying from Munda to Guadalcanal in a P-38 Lightning. Stewart fired a single long burst at the Betty and watched as what appeared to be thick black oil emerged from the bomber. However, the Japanese pilot managed to reach a cloud bank and Stewart could not find him despite a search in the mass. The pilot did not bother to report the incident. Later the wreckage of a Japanese Betty bomber was discovered in the vicinity of Stewart's aerial attack. Stewart decided to not say anything about the engagement with the Betty rather than receive a reprimand for not reporting the engagement in his initial post-mission report. *Courtesy of Dick Stewart.*

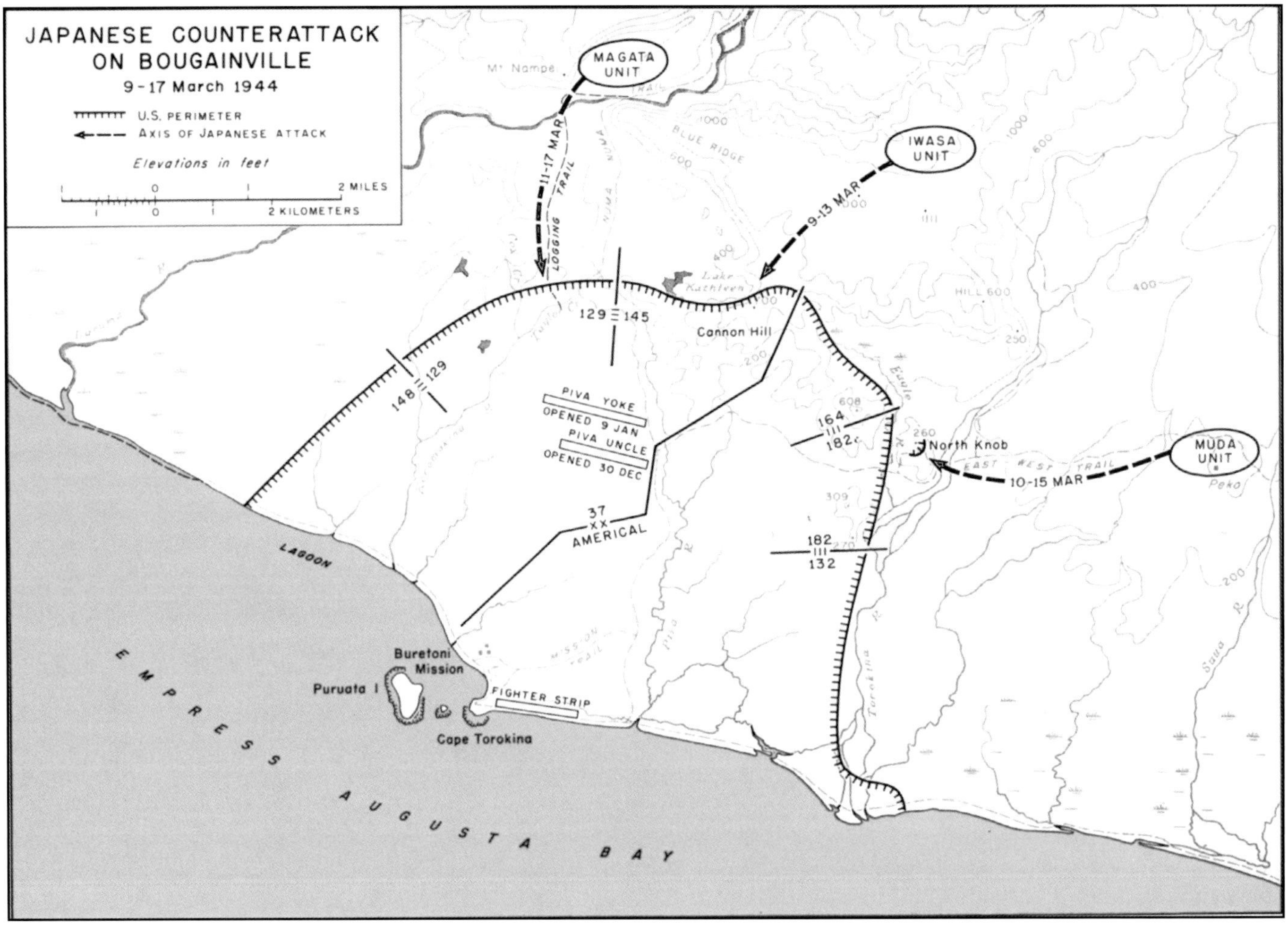

March 1944 Japanese counterattack on Bougainville. *Courtesy of the Center for Military History.*

Betty was simply a "desperation shot" by Stewart just prior to losing sight of the Japanese bomber. Stewart fired a single long burst at the Betty and watched as what appeared to be thick black oil emerged from the bomber. However, the Japanese pilot managed to reach the cloud bank and Stewart could not find him despite a search in the mass. A frustrated Stewart returned to Guadalcanal, where he asked his crew chief if the P-38 carried a gun camera. When the crew chief replied that his fighter did not carry a gun camera, which would have recorded the action, Stewart simply elected to not report the incident. Later the wreckage of a Japanese Betty bomber was discovered in the vicinity of Stewart's aerial attack. Stewart decided to not say anything about the engagement with the Betty rather than receive a reprimand for not reporting the engagement in his initial post-mission report.[26]

The incident with the Betty bomber was not the only time Stewart chose to ignore an incident in his report. During a P-38 mission over Bougainville, Stewart's plane was hit by Japanese anti-aircraft fire and a small piece of shrapnel embedded itself in the pilot's leg. He refused to tell anyone about his leg, which was very sore by the time he landed. Stewart removed his boot and his leg was covered in blood. A small hole in the leg indicated where he had been hit. Concern over being taken off flying status motivated Stewart to not report the wound and treat it himself. Fortunately the wound healed, but Stewart never received a Purple Heart medal, since he never reported the incident to medical authorities.[27]

One night on Guadalcanal squadron personnel were awakened by a series of tremendous explosions as an ammunition dump went up in flames. The dump was placed off limits as ordnance continued to detonate over the next couple of days. Approximately two weeks later, First Lieutenants Don Dessert and Bob Winslow sneaked into the dump along with Flight Officer Sanford "Sandy" Bauman. The trio entered a small building that was still standing and began looking for "treasures" among the ruins. Winslow opened a box to discover that it contained hand grenade detonators. He pulled the ring on a detonator out of curiosity. Upon realizing that he had just released the safety on the detonator, Winslow dropped it to the floor and put his foot on it. Dessert was bending over and looking at some boxes when the detonator exploded. Tiny pieces of the detonator flew out from underneath Winslow's boot and struck Dessert. Bauman picked up the story years later:

"Now what to do? Don had small slivers of junk in his legs and we were in forbidden territory. We got into our jeep and rushed back to camp, where our flight surgeon picked out the scrap metal one piece at a time."[28]

Detachment A of the 419th NFS, commanded by newly promoted Captain Al Lukas, operated from Bougainville when the Japanese launched their major counter attack on the island in March. The Japanese believed the Allied invasion at Empress Augusta Bay on November 1, 1943, was a diversion, and the main attack would occur somewhere else on the island. Despite the increasing flow of men and materiel onto the island and the development and construction of three air strips, the Japanese continued to believe this was a diversion. In early 1944 the Japanese finally recognized that the landing site was the main invasion point and began to gather their forces from all over the island for a counter attack. Bougainville consisted of thick jungle, with a high ridge running down the center of the island. Each end of the ridge nestled a mountain. It took considerable time for over 15,000 Japanese soldiers of the 6th and 17th Divisions to move on foot and converge for the thrust against the Allied positions. Allied intelligence reported the movement of Japanese troops all over the island, and Japanese prisoners surprisingly were very willing to divulge what they knew of the plans. Thus, by the end of February 1944 the Allies expected a Japanese counter attack sometime in the next couple of weeks.

The Japanese plan called for three separate forces to assault the Allied perimeter. The Iwasa Unit targeted the Pima airfields. Detachment A flew from Pima Uncle airfield and was located in the path of this arm of the assault force. The Masata Unit planned to seize the Torokina airfield along the coast and then assist the Iwasa Unit. The third force, the Muda Unit, aimed to attack from the east in order to divert forces and support the other two assaults. In mid-February the Allies withdrew the Fijian military outpost at Ibu following a major assault. This attack provided another indication that the Japanese forces were on the move. The Japanese estimated there were 30,000 Allied personnel on Bougainville. However, their intelligence estimates grossly underestimated the number of allied personnel, which actually totaled approximately 62,000. The American XIV Corps held the perimeter around the airfields with the Americal Division on the east and the 37th Division on the West protecting the Pima airfields.

The main Japanese thrust began on 8 March with an artillery barrage. Shells fell around the Pima airfields as aircraft were launched and withdrawn to New Georgia. Corporal Jesse Gardner, a 419th NFS Operations clerk, was assigned to Detachment A when the Japanese initiated their artillery barrage:

"We had just gone into Operations when shells began exploding across the road from us. I fell flat on the ground, grabbed my helmet, and jumped into the ditch in front of our tent".

Gardner and Sergeant Carl Reuben shared the ditch as shells continued to fall in the area. The two men straightened up to take a look out of the ditch. As they rose, another man yelled for them to "duck" and they quickly complied. Afterwards the man, Sergeant Marshall Peoples, from their squadron, informed them that he actually saw a large piece of shrapnel pass between their heads and bury itself in the bank of sand behind them. Skeptical at first, Gardner dug into the sand during a lull in the shelling and found a large jagged piece of shrapnel just as indicated by Peoples.[29]

Two 419th NFS pilots were airborne and returning from night fighter strip alerts when the barrage commenced. First Lieutenant Marvin Walker, returning from Green Island, landed his P-38 just prior to the shifting of the shelling to Pima Uncle. After a quick taxi he parked the P-38, dismounted, and dove into the ditch with Gardner and Reuben when the shelling commenced. Captain Lukas had pulled the night alert in the Treasury Islands and was returning to Bougainville at the same time that the shells were falling on the airstrip area. He landed in the midst of the Japanese barrage and made it safely to cover. Sergeant Cleve Wilborn, transportation NCO for Detachment A, lay in a trench near the airstrip as Walker and Lukas landed:

"Overhead we could hear the shells zooming and then a big explosion followed by pieces of shrapnel buzzing over our heads. One piece about 2 ½ inches long and ½ inch wide whipped by another guy's head. Had he been standing up, we would have had a casualty."[30]

Corporal Carl Ranck (center) was awarded the Purple Heart medal for a wound he received during the March 1944 Japanese counterattack on Bougainville. Major General Hubert Harmon is presenting the award. A second 419th NFS ground crew member, Sergeant Cleve Wilborn, also received a wound and was awarded the Purple Heart by a Marine Corps officer. Wilborn did not seek full medical treatment and always carried a small piece of shrapnel in his stomach. *Courtesy of Don Dessert.*

Around noon, dive bombers struck back at the Japanese artillery positions. Wilborn watched as the planes searched for and attacked their targets:

"We could see when they released their bombs. We could see the bombs going down, down, down, and then a terrific explosion. The boys were shouting at the top of their voices, 'Shoot high sheriff' and 'Let 'em have it!' We got a great thrill out of it. That afternoon, Detachment A received orders to dig in for a line of defense. We never dreamed anything like this would happen. We dug in about five yards apart. The boys were busy digging without any words at all. Captain Lukas and Lieutenant Winslow were in the center and they planned to direct our fire."[31]

First Lieutenant Dick Stewart and the other pilots prepared to defend their positions along with the enlisted men. Around 8:00 PM Marines informed the squadron that they heard the Japanese had broken through the American first line of defense. Stewart noted that they could hear the occasional voices of Japanese soldiers in the distance. During the night individual Japanese soldiers managed to infiltrate the lines, but entire units were held by the Americans. "The next morning our tents were full of holes and the nearest dead Japanese soldier was only 150 feet from our tent."[32]

The Japanese shelling continued the next day. Wilborn remembered:

"We were sitting in our tent talking and all of a sudden we heard a 'whiz' over our heads. An enemy shell hit approximately 200 yards from our tent. We did not wait for a special invitation to enter our fox hole. We

Left to right: Sergeant Willard "Wee Willie" Fellows and First Lieutenant Bob MacDonald. The 419th NFS lost MacDonald during a four ship P-38 Lightning attack on a Japanese gun position. With his right engine on fire, MacDonald maneuvered from the scene and made a successful bail out as confirmed by First Lieutenants Don Dessert and Dick Stewart. Word received later from a coast watcher indicated a Japanese patrol found MacDonald, who opted to shoot it out rather than surrender. MacDonald died in the exchange of gunfire, but unconfirmed reports indicate more than one Japanese soldier fell to his pistol in the process. *Courtesy of Don Dessert.*

stayed there for about 30 minutes while the 'whizzing' continued. The shelling continued, so the three of us slept in our fox hole that night. We could hear rifle fire and machine gun fire throughout the night."

That night Wilborn climbed out of his foxhole and sat down. A Japanese shell landed 20 feet away and a piece of shrapnel struck him in the stomach. Blood shot from his stomach. The 419th NFS flight surgeon, on the island with Detachment A, placed a piece of gauze over the wound and told Wilborn to apply direct pressure on it while he assisted others in the area also wounded in the attack. Wilborn never had the small piece of shrapnel removed from his stomach. The wound healed around the shrapnel and he carried it without any trouble after war. Wilborn hung his blood soaked T-shirt on a line next to his tent:

"A Marine captain came to our area later. He asked who owned the shirt. It was really bloody and I told him it was mine. Later he came back and gave me a Purple Heart."[33]

Allied forces launched a counter attack on 11 March. However, Japanese artillery and mortars continued to rain down on the Allied held areas for many days. Japanese infiltrators also continued to harass the air crews. On 13 March approximately 15 officers and NCOs of the 419th NFS were in the Detachment area when shelling commenced, and word arrived that the Japanese had broken through the defensive lines. Corporal Gardner was with this group:

"I sure was scared. We all were. After a wait of half an hour word came that all was secure, and we started walking up to the area…I had just crossed the parking strip in front of our diggings when a sniper opened up on us."

Later a second sniper fired upon Gardner, but someone evidently removed both snipers from being a threat.[34]

A second 419th NFS ground crew member was wounded on 18 March. Japanese shelling against the Allied airfield at Torokina commenced around 3:00 AM and was followed an hour later by a Japanese raider flying overhead despite the presence of American night fighters in the area. The bomb exploded

First Lieutenant Fran Parker. On May 2, 1944, Parker flew a special attack mission on Bougainville in cooperation with the Marines. Parker's P-38 carried a 125 pound experimental bomb developed locally by the Marines. One specific example of these experimental bombs contained 400 rounds of .45 caliber ammunition, 50 rounds of .30 caliber ammunition, 40 pounds of high explosives, eight signal flares of different colors, smoke bombs, and other various types of ordnance. The purpose of this type of weapon involved "shock and awe" on the Japanese forces. When the bomb exploded, it was hoped that the bullets would fire into various directions, the high explosives would go off, the signal flare bombs would ignite into their assorted colors, and the smoke would cover the area. Parker flew two of these experimental missions, which did not produce the results hoped for by the Marines. *Courtesy of Leslee Lukas Mays.*

Second Lieutenant Bernard Knott (photo taken after his promotion to captain) flew an incendiary cluster bomb mission with First Lieutenant D. C. Bray over Bougainville. Each P-38 Lightning carried a 500 pound incendiary cluster bomb. The pilots dropped their ordnance on the Japanese held Inus Plantation bivouac and supply area, where the bomblets dispersed over a wide area as they ignited small fires, as well as a larger blaze on a building. Knott later survived a P-61 Black Widow crash at sea in the Philippines during July 1945. *Courtesy of AFHRA.*

approximately 50 yards from the 419th NFS tents, killing a sailor and wounding five other sailors. A piece of the shrapnel struck Corporal Carl Ranck of the 419th NFS in the arm. Ranck was not badly hurt and his arm was treated in the local hospital. He received a Purple Heart medal for the wound soon afterward in a ceremony, along with some of the pilots who were being awarded Air Medals for their combat missions. Although Ranck, Stewart, and Wilborn received wounds while overseas, Ranck was the only one to receive official medical treatment. Stewart and Wilborn were fortunate that their wounds never became infected and worsened due to their decisions to not report them.[35] Within days, the Allied counter attack forced the Japanese forces back into the jungle and eliminated the few soldiers left behind.

One of the saddest incidents of the war for the 419th NFS personnel occurred the same day that Ranck was wounded. At 11:45 AM First Lieutenant Marvin "Marvelous Marv" Walker took off from Bougainville in a P-38 and was never seen again. Rumors floating back to Guadalcanal indicated that he had been shot down and beheaded by the Japanese forces still on Green Island. However, this rumor was far from the truth. Walker had requested external fuel tanks be added to the P-38, which confused the ground crews, since a flight to Guadalcanal did not require additional fuel than what was carried internally within the fighter. Walker disappeared and was never seen again. Piecing together the story later, the squadron ascertained that Walker decided to attempt a flight to Tahiti, located far to the east of the Solomon Islands and beyond the range of a P-38 even with external fuel tanks. Reports surfaced of a P-38 crash landing at sea near an island and immediately sinking beneath the surface. Captain Howard Daniel went to investigate the report. His findings confirmed that the crash was probably Walker and his P-38. The exact location of the crash was never discovered, nor the remains of the pilot recovered. No one fully understood why Walker would attempt to depart the Solomons. Some believed he suffered from homesickness, while others believed that he had heard tales of Tahitian women and flew off on an unauthorized vacation. Whatever the reason, the mental strain of the Japanese attacks on Bougainville during the past few days may have contributed to his state of mind when he departed.[36]

The 419th NFS lost another pilot during a combat operation on 18 April. First Lieutenants Don Dessert, Bob MacDonald, Robert LaCroix, and Dick Stewart were on a patrol to locate and destroy a Japanese coastal artillery piece on Buka Island, located off the northwestern tip of Bougainville. The aircraft made two passes on the artillery emplacement and dropped their ordnance. However, MacDonald's P-38 took a hit from anti-aircraft fire on his second pass. With his right engine on fire, MacDonald maneuvered from the scene and made a successful bail out as confirmed by Dessert and Stewart. MacDonald never made it off the island. The three P-38 pilots flew cover over the location where MacDonald landed in his parachute and were later relieved by New Zealand and U.S. Navy fighters. Two of the New Zealand aircraft and two Navy planes were shot down as they flew cover for MacDonald in order for him to try to make it to the coast for a rescue. A coast watcher later reported that MacDonald and the other four pilots were found by the local inhabitants and taken to a village. Someone in the village reported their existence to the Japanese, who then arrived to seize them. The New Zealand pilots managed to escape from the village, but for whatever reason MacDonald opted to shoot it out with the Japanese patrol. MacDonald died in the exchange of gunfire, but unconfirmed reports indicate more than one Japanese soldier fell to his pistol in the process. The two Navy pilots were captured and reportedly beheaded.[37]

Throughout the remainder of April and May, the 419th NFS pilots continued their harassing P-38 missions over Bougainville and other islands in the northern Solomons. In May the squadron flew 91 sorties ranging from solo missions to larger operations consisting of two, three, and even four P-38s. The one major difference between the harassing missions at the opening of 1944 and those in the spring of 1944 involved weapons usage. The early missions utilized the P-38s' machine guns and cannon to strafe Japanese positions, while the latter operations included 500 and 1,000 pound general purpose, cluster, and incendiary bombs. A general purpose bomb explodes on contact with a target as its casing brakes apart and spreads in all directions as shrapnel. Cluster bombs carry many smaller bomblets within their casings. As it falls, the casing opens and bomblets spread across a wider area. Incendiary bombs are designed to start fires when they explode. General purpose bombs are best utilized on physical objects, such as bridges, buildings, or other types of infrastructure, while cluster bombs are more effective when employed against personnel on the ground. Incendiary bombs are effective on anything that will burn, but especially fuel and ammunition dumps.

First Lieutenant D. C. Bray and Second Lieutenant Bernard Knott flew an example of a mission utilizing a combination of these weapons on 2 May. Bray and Knott took off in P-38s from Torokina at 2:30 PM. Each P-38 carried a 500 pound incendiary cluster bomb. At 2:50 PM the pilots dropped their ordnance on the Japanese held

Northrop test pilot John Meyers (standing on the right). Meyers arrived on Guadalcanal to demonstrate the capabilities of the P-61 Black Widow and help the 419th NFS pilots transition to their new aircraft. 419th NFS veterans related that Meyers could do things with the P-61 that still amazed them years after the war. *Courtesy of Don Dessert.*

Pratt and Whitney technical representative (tech rep) Marshall Holbrooke. Pratt and Whitney manufactured the engines utilized on the P-61 Black Widow. Holbrooke taught the 419th NFS maintenance crews how to service the Black Widow's engines. *Courtesy of Don Dessert.*

Left to right: Northrop technical representative (tech rep) M. Scott Johnson and Northrop test pilot John Meyers visiting Detachment A of the 419th NFS on Bougainville. Scott was on hand for the arrival of the P-61 Black Widows and helped supervise their assembly. Meyers arrived a few days later to begin training the pilots. *Courtesy of AFHRA.*

Inus Plantation bivouac and supply area. The bomblets dispersed over a wide area as they ignited small fires, as well as a larger blaze on a building. The P-38 pilots returned to Torokina and landed at 3:30 PM.[38] First Lieutenant Dorval Brown and Second Lieutenant James Lane flew an example of a general purpose bomb mission on 9 May. The pair departed Torokina at 10:00 AM in P-38s and headed for Numa Numa Harbor and Plantation. Upon arriving at the target area 30 minutes later, the pair launched a diving attack with the single 500 pound general purpose bomb carried by each P-38. One bomb fell short of the target and landed in the Plantation area as a dud. The second bomb exploded in a supply area, but the pilots were not able to observe the results, as they sped away from the area before anti-aircraft fire could be directed at them. The pilots returned to Torokina at 11:30 AM.[39]

On 2 May the 419th NFS cooperated with the Marines in the employment of an experimental bomb. First Lieutenant Francis Parker took off from Torokina at 3:45 PM and joined a Marine Corps pilot flying a Douglas SBD Dauntless for a raid on Kieta gardens, near Kieta Airfield. Parker's P-38 carried a 125 pound experimental bomb developed locally by the Marines. One specific example of these experimental bombs contained 400 rounds of .45 caliber ammunition, 50 rounds of .30 caliber ammunition, 40 pounds of high explosives, eight signal flares of different colors, smoke bombs, and other various types of ordnance. The purpose of this type of weapon involved "shock and awe" on the Japanese forces. When the bomb exploded, it was hoped that the bullets would fire various directions, the high explosives would go off, the signal flare bombs would ignite into their assorted colors, and the smoke would cover the area. A 419th NFS pilot tested the bomb in cooperation with the Marines on 22 April. A COMAIRSOPAC report noted "Peering through rain, the pilot of the P-38 concerned saw a brilliant flash, flares shooting out in all directions, and a cloud of smoke." Parker's 2 May mission was not as successful as the one in April. At 4:04 PM Parker approached the target at 400 feet and dropped the experimental bomb, which failed to react as its predecessor. Parker reported the bomb burned in a ten foot perimeter for a couple minutes before fizzling out. The Marine Corps pilot intended to drop two fuel tanks on the target after the bomb exploded in order to encourage the spread of the destruction. With the failure of the bomb both pilots returned to Torokina and landed at 4:45 PM.[40]

Unassembled Northrop P-61 Black Widows arrive at Guadalcanal aboard a U.S. Navy vessel for the 419th NFS. *Courtesy of Garry Pape via Jeff Kolln.*

Moving the 419th NFS P-61 Black Widows to shore at Guadalcanal. *Courtesy of Garry Pape via Jeff Kolln.*

Lifting new 419th NFS P-61 Black Widows from a U.S. Navy vessel. *Courtesy of Garry Pape via Jeff Kolln.*

The 38th Service Squadron begins the task of helping assemble the P-61 Black Widows destined for the 419th NFS.
Courtesy of Garry Pape via Jeff Kolln.

Three days later, on 5 May First Lieutenant Brown and Second Lieutenant Lane flew another mission with one of the Marine Corps' experimental bombs. The pilots took off in P-38s at 10:00 AM from Torokina and flew to the Japanese held villages of Melelup and Tumuri. Brown and Lane arrived at their targets by 10:30 AM and initiated a dive bombing attack from 7,000 feet. They released their bombs at 5,000 feet. Lane carried a single 500 pound incendiary cluster bomb that ignited many small fires in the village of Melelup. Brown's P-38 had been armed with a 500 pound incendiary cluster bomb and one of the smaller experimental bombs. The intention of this combination was to utilize the incendiary cluster bombs to help ignite and spread the ordnance contained inside the experimental ordnance. Brown dropped both of the bombs on Tumuri village and noted that the combination started numerous small fires, but he had nothing spectacular to report. Following the attack, the pilots returned to Torokina and landed at 11:00 AM.[41] Squadron reports do not record any further use of the Marine Corps' experimental bombs. The 419th NFS continued to fly harassing missions over Bougainville for the remainder of the month.

The "day of days" finally arrived on 1 May. Word reached the 419th NFS headquarters that a vessel off Guadalcanal carried unassembled Northrop P-61 Black Widows. After six months of disappointment and being forced to fly intruder and harassing missions in P-38s and P-70s, the 419th NFS finally had an aircraft designed for night interception of enemy aircraft. Although their P-38 missions over Bougainville would continue for the remainder of the month as discussed earlier, the ground crews utilized May to assemble the Black Widows. Two corporate representatives arrived with the aircraft: John Meyers, the primary test pilot of the P-61; and Marshall Holbrooke, who provided advice and training on the maintenance of the Pratt and Whitney engines. Meyers demonstrated what the P-61 could accomplish and began instructing the pilots on the flying characteristics of the new fighter.

The Northrop P-61 Black Widow emerged from the drawing board as a large fighter designed for three

Maintenance of a 419th NFS P-61 Black Widow 42-5507 #312 on Guadalcanal. *Courtesy of Don Dessert.*

crewmen. The plane, over 49 feet in length, was powered by two Pratt & Whitney R-2800-65 Double Wasp 18-cylinder radial engines. It could fly over 1,350 miles at over 220 miles per hour. The pilot's cockpit was located just behind the nose, which carried the radar receiving antenna. The gunner sat above and behind the pilot. A turret housing four .50 caliber machine guns was mounted just behind the gunner's cockpit. The positioning of the guns permitted them to be fired without their flash blinding the gunner at night. Later models of the P-61A and early models of the P-61B did not carry machine gun turrets due to trouble with their operation and a shortage of gun turrets. The P-61 utilized the same gun turret mechanism as the Boeing B-29 Superfortress bombers, which had a higher priority for receiving them. Later models of the P-61B did carry the machine gun turret. P-61s without machine gun turrets did not fly with gunners, and their only armament were the pilot operated 20 mm cannon. Each P-61 carried four 20 mm cannons in a tub located on the belly of the aircraft which were the primary responsibility of the pilot to operate. The R/O worked from a unique compartment at the rear of the P-61. During flight the R/O faced forward, but could turn in his seat to peer out the Plexiglas located behind him. The R/O operated the new SCR-720 radar system with an effective range of around five miles when locating aircraft. However, this distance could increase or decrease, depending upon various conditions. In another unique design feature, any member of the crew could operate the .50 caliber machine gun turret. Thus, if pursued by a Japanese fighter, the gunner could turn the .50 caliber machine guns to the R/O, who was in a better position to view the enemy plane and engage it. Although this would be an extremely rare event, 419th gunners did transfer the fire control of the .50 caliber machine guns to their R/Os on at least two occasions, and these will be discussed later. The plane could carry 200 rounds of 20 mm ammunition for each gun and 560 rounds of .50 caliber ammunition for each machine gun.[42]

By 12 May some of the 419th NFS pilots were sufficiently oriented with the P-61 to begin test hops in the plane. Major Emerson Barker, the squadron commander, made the first flight, and was soon followed by the other pilots. Each praised the P-61 for its flying characteristics. First Lieutenant Don Dessert noted that the Black Widow "was very agile for its size and a very large aircraft for a single pilot. Its maneuverability was good, and it could turn with the smaller P-38 without stalling."[43] Every pilot looked forward to ambushing Japanese aircraft with their new night fighter.

While the training continued the 419th NFS lost another P-38. On 19 May First Lieutenant Dudley Bray crash landed a P-38 at Bougainville, totaling the aircraft. The pilot walked away from the accident. On 12 June the squadron lost its first P-61 Black Widow. Major Emerson Barker (pilot), along with Captain Robert Signet (R/O) and Staff Sergeant Melvin Masek (gunner), departed Guadalcanal in a P-61 for a night training mission with five other P-61s. Barker's crew never returned. Barker was practicing P-61 dog fighting with First Lieutenant Dick Stewart (pilot) and his R/O, Flight Officer Jim

Another view of 419th NFS P-61 Black Widow 42-5508 #313 flying with only one engine during pilot transition training off Guadalcanal. *Courtesy of Don Dessert.*

Allen. The pilots alternated between hunter and being the hunted. The P-61 designated as the hunter flew local patterns until the R/O made contact with the plane being hunted. At that point, the hunter would maneuver for an intercept and then allow the other P-61 to disappear in order to renew the session. Allen remembered that he and Stewart intercepted Barker and his crew and then permitted them to disappear in order to repeat the exercise. Barker could not be found again. Stewart noted that he saw a brief flash on the water, and this could have been an exploding aircraft.[44] Crash boats and a sub-chaser immediately went out to search for the missing crew. They found an oil slick, but no sign of the crew members or the aircraft. Captain Al Lukas assumed acting command of the squadron as the personnel mourned the loss of their second squadron commander to an accident. Less than a week later, 13th Air Force assigned Major Andrew Bing as the new 419th NFS commander. Bing served as a P-38 day pilot in the North African campaign in 1942. The new commander did not adapt well to flying at night and requested a transfer two weeks later. He was replaced by Major Joseph Shulmistras.

On 27 June Captain Lukas led the first group of P-61s from Guadalcanal to Nadzab, on the eastern coast of New Guinea. Following more training at Nadzab, these pilots, as Detachment A, departed for Noemfoor Island, off the northern coast of New Guinea. With the Solomon Islands secured and Rabaul bypassed to wither on the vine, cut off from Japan, the 13th Air Force turned westward to support Operation Cartwheel—General MacArthur's movement across northern New Guinea in the advance toward the Philippines. The remainder of the unit headed to Los Negros, off the northeast coast of New Guinea, and then would establish operations on Middleburg Island. The departure of the 419th NFS from the Solomon Islands left the Marine Corps' VMF(N)-531 as the only night fighter squadron in the area. The Marine Ventura night fighters flew their last combat mission in the Solomons on 14 July and departed for Hawaii on 2 August. With the arrival of the 419th NFS' two detachments in New Guinea, the 13th Air Force's only night fighter squadron was finally entering the war with a true night fighting aircraft.

P-61 Black Widow 42-5506 #311 piloted by 419th NFS commander Major Emerson Barker—the first 13th Air Force P-61 Black Widow to become airborne in the Pacific. Barker, an original Detachment B, 6th NFS pilot, received credit for a probable kill of a Japanese Mitsubishi G4M Betty bomber on August 13, 1943. He disappeared without a trace during a P-61 training mission in June 1944. First Lieutenant Al Lukas assumed temporary command of the 419th NFS until the arrival of Major Andrew Bing, who remained only two weeks with the night fighters before departing. *Courtesy of Don Dessert.*

The first flight of a 13th Air Force P-61 Black Widow begins with short taxi to the air strip. *Courtesy of Don Dessert.*

419th NFS P-61 Black Widow 42-5508 #313 piloted by Captain Al Lukas during aircraft transition training off Guadalcanal. Note that both propellers are not spinning. Northrop test pilot John Meyers demonstrated and then taught the 419th NFS pilots how to land a P-61 without engines in an emergency. *Courtesy of Leslee Lukas Mays.*

The 38th Service Squadron helping assemble the P-61 Black Widows destined for the 419th NFS. *Courtesy of Garry Pape via Jeff Kolln.*

419th NFS P-61Black Widow 42-5505 #310 *Midnite Marjorie: Luke's London Lovely*. Captain Al Lukas claimed #310 as his "personal" aircraft, adding an altered version of his wife's nickname from his P-38 Lightning. Marjorie Kingsley, Lukas' wife, was a famous English singer during the World War II era, often performing with the Harry Roy orchestra and later with the British version of the USO. Lukas was flying #310 when he downed a Japanese Mitsubishi Ki-21 Sally bomber off Noemfoor three months later. *Courtesy of Leslee Lukas Mays.*

419th NFS P-61 Black Widow 42-5508 #313 piloted by Captain Al Lukas during aircraft transition training off Guadalcanal. Note that the right engine has been shut off to give the pilot practice at either restarting or flying on one engine. Also note the horizontal bar at the rear of the Radar Observer's nacelle. High speed dives tended to result in implosions of the nacelle and the bar helped to prevent this problem. *Courtesy of Leslee Lukas Mays.*

4

NOEMFOOR

"The Dinah went spiraling down in flames from 11,000 feet until it disappeared below the overcast, which was 2,000 feet below us."

Second Lieutenant James Schroth
pilot, 419th NFS

In June 1944, Washington, D.C., approved the establishment of the Far East Air Forces (Provisional), better known as FEAF (P). The new organization should not be confused with the Far East Air Force (FEAF), which commanded American air units at the opening of war in the Pacific in 1941-1942 and evolved into the 5th Air Force. FEAF (P), under the leadership of General George Kenney, comprised the 5th and 13th Air Forces and became a permanent organization as the Far East Air Forces (FEAF) in August 1944. Under Kenney's direction, the 5th Air Force assumed the air offensive lead in General Douglas MacArthur's campaign across New Guinea toward the Philippines. The 13th Air Force, the smaller of the two organizations, provided a more supporting role to 5th Air Force.

July 1944 opened with the 419th NFS divided into two detachments. Detachment A, commanded by Captain Al Lukas, with First Lieutenant John Laffey

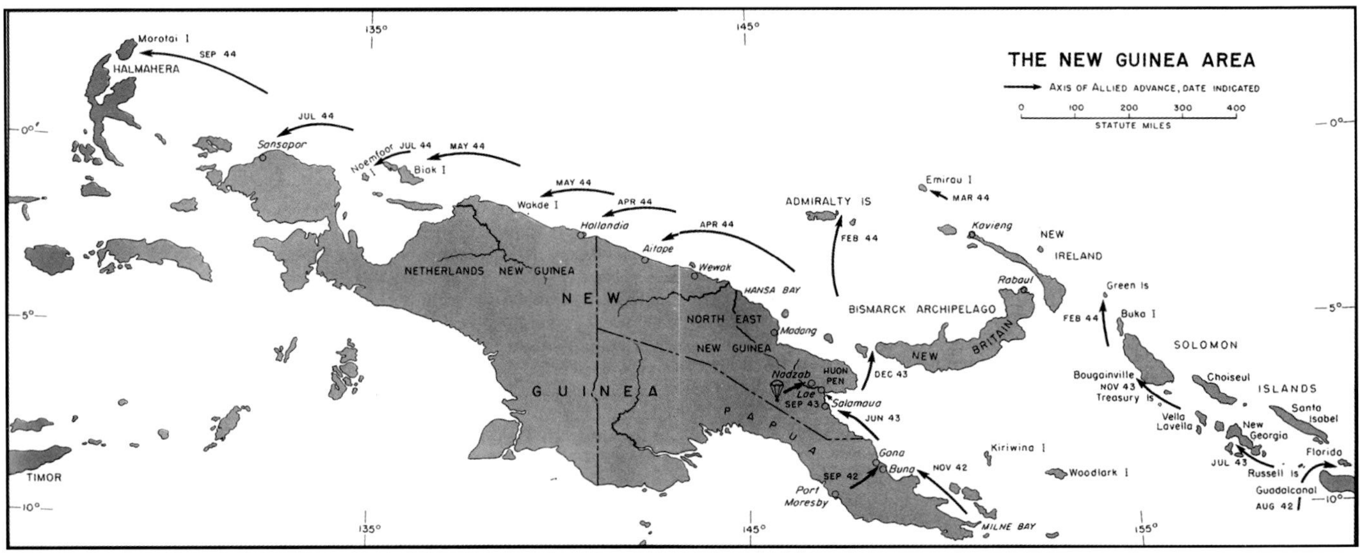

Map of New Guinea and the Solomon Islands. The 13th Air Force night fighters (Detachment B of the 6th NFS and then the 419th NFS after November 1943) began at Guadalcanal, in the lower right of the map, defended the Solomon Islands, shifted to the north coast of New Guinea, and then followed the Allied offensive westward to Morotai, from which they then deployed northward to the Philippines. *Courtesy of the U.S. Army.*

One detachment of the 419th NFS conducted additional P-61 training at Nadzab, on New Guinea, after departing Guadalcanal. Morale crumbled at Nadzab due to the quality of food and lack of off duty activities that were more abundant on Guadalcanal. *Courtesy of Leslee Lukas Mays*

as operations officer, prepared for a move to Nadzab, New Guinea, headquarters of the 5th Air Force. Plans called for Detachment A to be temporarily attached to the 5th Air Force, while Detachment B and the 419th NFS Headquarters section remained with the 13th Air Force. Detachment B, with Captain Howard Daniel in command and Captain Charles Van Bibber, formally of Detachment B 6th NFS, serving as the operations officer, continued its training at Guadalcanal and would soon transfer its aircraft to 13th Air Force Headquarters at Los Negros, in the Admiralty Islands, located northeast of New Guinea.

Twenty-two C-47 transport planes were required to move Detachment A's personnel and equipment to Nadzab. The unit established operations with its six Northrop P-61 Black Widows at Strip #3. Aircrews utilized the time to hone their air-to-air and air-to-ground gunner skills and practiced radar assisted intercepts. Ground personnel continued to develop their maintenance skills on the new P-61 aircraft and pulled mundane duties, such as guard mounts and mess hall assistance.

Morale of the men dipped at Nadzab. Many complained of the "monotonous meals" at the mess hall. Breakfast alternated between pancakes and coffee one morning and powdered eggs and coffee the next day. Fresh meat and vegetables were unavailable, resulting in a rotation of bully beef, Vienna sausages, canned stew, chili con carne, and baked beans for main courses. Bread and jam were plentiful, and many men practically lived off the combination for the month. Five men came down with dengue fever during July. These cases represented a greater total than the entire 419th NFS had suffered between November 1943 and June 1944. The movies shown in the outdoor theaters tended to be older films that the detachment personnel had seen numerous times. An occasional softball game and a radio tended to be the only non-work distractions for the personnel. The men of Detachment A were ready when they received orders to establish combat operations on Noemfoor, an island off the northwestern coast of New Guinea.

Allied forces under General Douglas MacArthur were hopping along the northern coast of New Guinea toward the Philippine Islands, bypassing many Japanese garrisons along their path. After establishing an advance base at Biak Island, Allied planners launched an assault on Noemfoor Island. Noemfoor lay in Gellvink Bay, halfway between Allied occupied Biak and the Japanese occupied Vogelkop Peninsula of western New Guinea. Capture of the island could provide the Allies with a second base in the area and help deny the use of the island by Japanese forces attempting to reinforce their besieged garrison on Biak. Allied forces on Noemfoor could then support the next series of assaults destined for the Vogelkop Peninsula. Noemfoor, a circular coral and limestone island, is approximately fifteen miles (north-south) by twelve miles (east-west) in size. Three Japanese airfields of varying degrees of completion were located on the island and could be improved for Allied use. The 158th Infantry Regiment made the initial assault of Noemfoor on July 2, 1944, followed by airborne landings by the 503rd Parachute Infantry during the next two days. Throughout July Noemfoor held the distinction of being the most forward Allied base in New Guinea, and operations to clear the island of Japanese resistance would continue until 31 August.[1]

Japanese night intruders flying from bases at Ceram, located off the southwest coast of New Guinea, staged through Moemi Airdrome to make nine sorties in five days over Noemfoor beginning on July 4, 1944. Other Japanese airfields in the area included Halmahera, Amboina, and Boeroe, with bases at Mindanao and Palaus further away but within bomber range. All of the July Japanese intruder raids on Noemfoor were ineffective,

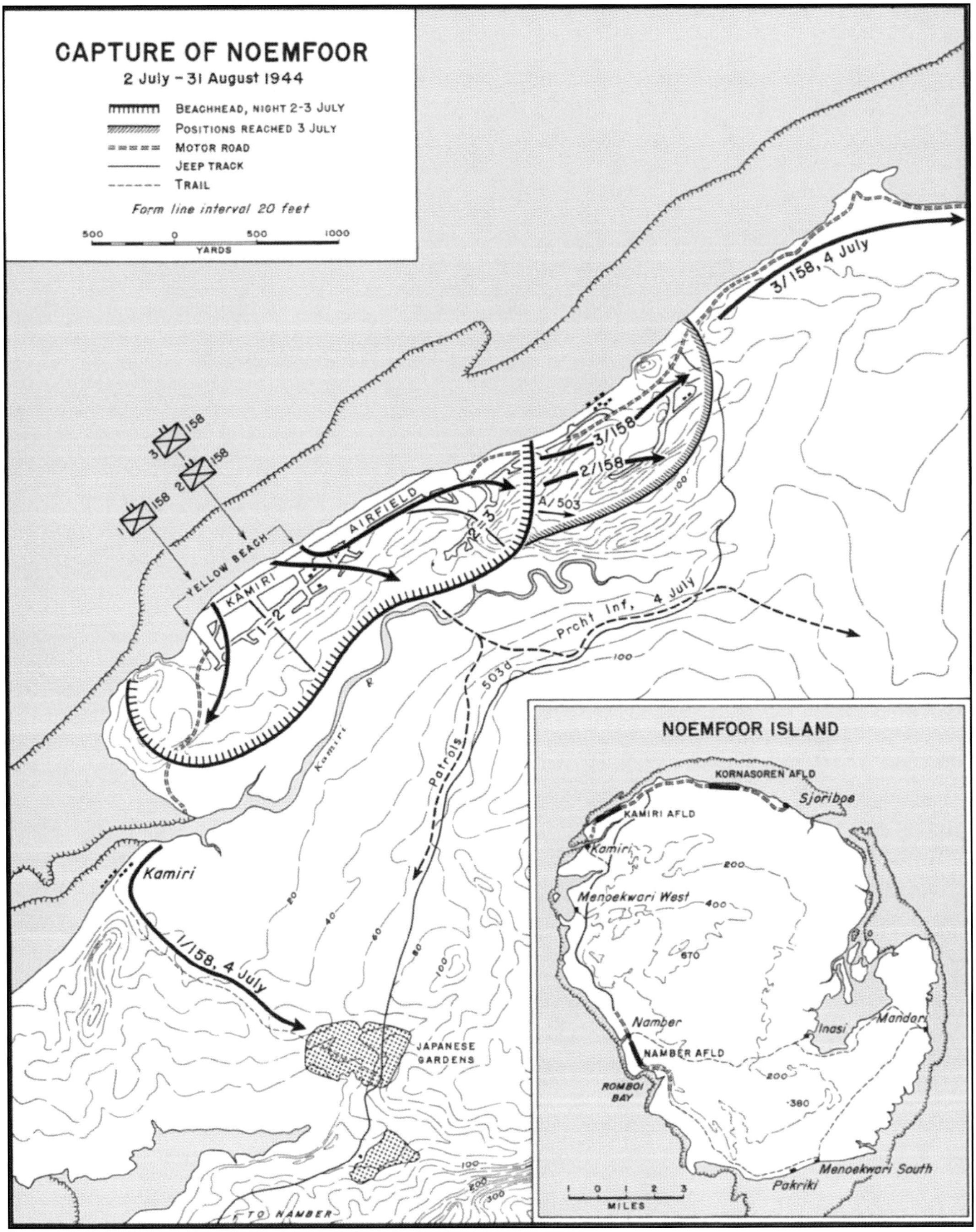

Noemfoor. *Courtesy of the Center for Military History.*

but illustrated the need for American night fighters on the island. Headquarters, 5th Air Force, tapped Detachment A of the 419th NFS to provide local nocturnal protection over Noemfoor, as well as coverage for the upcoming Allied landings at Sansapor, on the Vogelkop Peninsula, at the end of the month. The Black Widows of the 419th NFS were eager to weave their webs between Biak and the Vogelkop Peninsula and dare Japanese intruders to test them.

The first Douglas C-47 Skytrains (the military version of the civilian Douglas DC-3 passenger plane) carrying Detachment A's personnel and equipment departed Nadzab on 25 July. Of the 11 transport planes that took off that morning, three aircraft reported weather related problems and returned to Nadzab. The other eight planes did not encounter any trouble during the flight. It may have been a coincidence, but it seems a well known movie actress was making a USO appearance that evening at Nadzab. All of the C-47s carrying Detachment A personnel and equipment landed at Biak over the next couple of days, where the unit transferred to Noemfoor in landing craft.

Detachment A arrived at Kamiri Strip, on Noemfoor, to discover their bivouac area would be located in "virginal tropical jungle" approximately three miles from their aircraft at the airfield. Clearing the jungle for the pitching of pyramidal tents involved back breaking work with axes and machetes. Local authorities gave the detachment six old jeeps and a single 250 gallon water trailer. All services were located miles from the bivouac area, adding to the drooping morale of the men. The situation improved when a bulldozer driver was persuaded by a "private deal" to clear the jungle around the tents. This allowed the men to increase the distances between their tents. Japanese soldiers remained on Noemfoor, necessitating detachment personnel to maintain a constant vigil with guard mounts in the bivouac area and their flight line. Rain and muddy conditions added to the long list of problems faced by the new arrivals. The rain would contribute to the death of a pilot in August.

Operationally, maintenance personnel were baffled by the temperamental SCR 720 radar sets on the P-61 Black Widows. The crews worked day and night to keep them operational, only to have the sets fail when

419th NFS P-61 Black Widow 42-39475 #338. The Black Widow carried four .50 caliber machine guns in a top turret and four 20mm cannons in its belly. The P-61 proved to be a very lethal adversary when all of its gun systems were firing at once. Many P-61 kills were made with just two or three short bursts of its eight guns. Later production "A" models and early production "B" models of the P-61 did not include the machine gun turret due to a more pressing need to mount them on Boeing B-29 Superfortresses being assembled for the air offensive against the Japanese homeland. P-61s without turrets were flown by crews that included only the pilot and radar observer. *Courtesy of Bob Fowble.*

Four of the Ground Control Intercept (GCI) personnel attached to the 419th NFS. The GCI controllers manned the ground based radar system that directed the night fighters. The GCI controllers directed a P-61 Black Widow to a position near an incoming bogey until the P-61's onboard Radar Observer could make contact with his own radar system. Operating the early radar systems and directing the Black Widows via altitude and directional headings to positions that would lead to intercepts proved to be very challenging. Captain Al Lukas of the 419th NFS noted that a successful night fighter intercept demanded skillful GCI controllers working in close cooperation with the P-61 pilot and radar observer. *Courtesy of Don Dessert.*

airborne. Master Sergeant William "Bill" McConnell, the maintenance line chief, remarked how the SCR 720 demanded considerable power to operate. "The radar would burn up the generators on board the P-61s and they were not easy to replace." Radar and communications failures would contribute to the loss of kills against bogeys at Noemfoor and other locations the 419th NFS served.[2]

An important key to successful night fighting involved the coordination of the P-61 pilot, his Radar Observer (R/O), and the Ground Controlled Interception (GCI) operator, or "controller." A GCI station was established near a Fighter Control Center and frequently operated two radar systems: one for day and one for night operations. The night fighter GCI controllers who worked with the 419th NFS were formally attached to the unit and traveled with them from station to station. The obvious advantage involved the development of a relationship between the GCI controller and the night fighter pilots and R/Os with whom he worked. When an unidentified aircraft approached the area without the proper Identification Friend or Foe (IFF) signal, the GCI would alert the patrolling night fighter or scramble a P-61 sitting on alert status. The GCI operator then assumed control of the P-61 and provided him with a vector to the incoming target. The object was to direct the P-61 to a location where the Black Widow's R/O could acquire a fix on the bogey using the onboard SCR 720 radar system. Then, the R/O would direct his pilot to a location where the crew could make visual contact with the unidentified aircraft. This was important, because the bogey could be a friendly aircraft with communications and/or IFF problems.

A night fighter on patrol would fly on the outside of an Anti-Aircraft (AA) circle established by the local gun battery. The imaginary circle, known by many pilots as the "bathtub," encompassed an area around an airbase where the AA battery would automatically open fire on any unidentified aircraft. They would not necessarily discontinue firing if a friendly night fighter entered the AA circle in pursuit of a bogey. Thus, most night fighter pilots heeded the warnings of the GCI controller when approaching the AA circle, even if chasing a Japanese intruder. Pilots who had visual contact with a bogey that was entering the AA circle could request the GCI controller to order a "hold fire" to the AA battery and permit him a continuation of the pursuit. The GCI controller could also order a "hold fire" to the AA battery if a P-61 was being scrambled to intercept an unidentified aircraft and then lift the order once the Black Widow was clear of the AA circle.

Allied pilots radioed their call signs whenever approaching an airfield to ensure they would not be misidentified as Japanese planes. This was especially important for night fighters landing after sundown. John Radzietta noted:

"Each pilot had a specific call sign that included a prefix followed by two numbers. Our commander was 'Adder One-Zero' and Captain Daniel was 'Adder One-One.' While I was flying with the commander he called in

First Lieutenant William Roney (seated at the bottom right). Roney, along with First Lieutenant William Zint (radar observer) and Staff Sergeant Robert McCall (gunner), flew as a team when their P-61 Black Widow suffered an engine power failure immediately upon take off and dropped into the ocean during July 1944. Roney and Zint died in the impact, while McCall was thrown clear of the wreckage. McCall managed to separate himself from his seat and rose to the surface despite the pain of a broken and severely gashed leg, a dislocated hip, fractured ribs, a punctured lung, and numerous cuts and bruises. His damaged Mae West survival vest kept him afloat until a rescue craft reached him. This was McCall's second crash survival in as many days. *Courtesy of Leslee Lukas Mays.*

'One-Zero approaching from the north' while attempting a landing. I quickly called the control tower from my R/O compartment and added, 'Correction tower, this is Adder One-Zero approaching from the north.' I did not want to get shot down for being mistaken as a Japanese Mitsubishi Zero."

Detachment A commenced combat operations from Noemfoor on July 28, 1944, with six P-61s and a P-70 they inherited on 11 July for administrative and supply flights. Immediately, the detachment provided night fighter cover for the American landings on Sansapor, as well as over Noemfoor itself. Three days later, Detachment A had been reduced to 50 percent of its combat strength following three successive days of P-61 accidents.

On July 28, 1944, First Lieutenant Herman Manke returned to Kamiri Strip from a local P-61 training hop at 4:45 PM. Air traffic controllers at Kamiri noticed that his nose wheel had failed to extend from the aircraft as he approached the field. The air traffic control tower flashed a red light to warn the pilot that there was a problem and he should not attempt a landing. However, Manke did not see the light and continued his approach, making a normal landing unaware of the predicament. The P-61's nose collapsed into the coral as its air speed dropped at the far side of the strip. Manke was not injured, but the P-61 suffered a severely damaged nose section and props.

The next night, Captain Lukas led a 4-ship flight of P-61s to provide a dusk patrol over the naval task force steaming to Sansapor. American forces planned to swarm ashore and establish a beachhead the next morning. After an uneventful patrol, the P-61s returned to Noemfoor at 8:30 PM. Lukas allowed the other pilots to land their Black Widows first and then followed them onto Kamiri Strip. Upon landing, Lukas struck the cab of a Royal Australian Air Force (RAAF) fuel truck parked in the dark alongside the runway. The impact sheared off the

First Lieutenant William Zint (second from the left). Zint, a radar observer, died along with his pilot, First Lieutenant William Roney, in a crash of a P-61 Black Widow in July 1944. To Zint's immediate right in the photo is Captain Dick Stewart holding the squadron mascot, *Old Salty Dog*. Captain Don Dessert sits immediately behind Stewart. *Courtesy of Leslee Lukas Mays.*

truck's cab and damaged the left wing of the P-61. The driver had stalled the truck and left it parked beside the runway. He did not inform the control tower, nor place warning lights on the vehicle. Fortunately, no one was injured in this potentially disastrous accident.

The third accident occurred on July 30, 1944, the night after Lukas hit the fuel truck. Three P-61s departed Kamiri airstrip at 5:45 PM to provide dusk night fighter coverage over the invasion fleet at Sansapor. First Lieutenant William Roney piloted the lead P-61 with First Lieutenant William "Bill" Zint as his R/O and Staff Sergeant Robert McCall riding shotgun as the gunner. The P-61 suffered an engine power failure immediately upon take off and dropped into the ocean. The plane disintegrated upon crashing into the water. Lieutenants Roney and Zint died in the impact. McCall was thrown clear of the wreckage and found himself still strapped into his metal seat while underwater. He managed to separate himself and rose to the surface, despite the pain of a broken and severely gashed leg, a dislocated hip, fractured ribs, a punctured lung, and numerous cuts and bruises. His damaged Mae West survival vest kept him

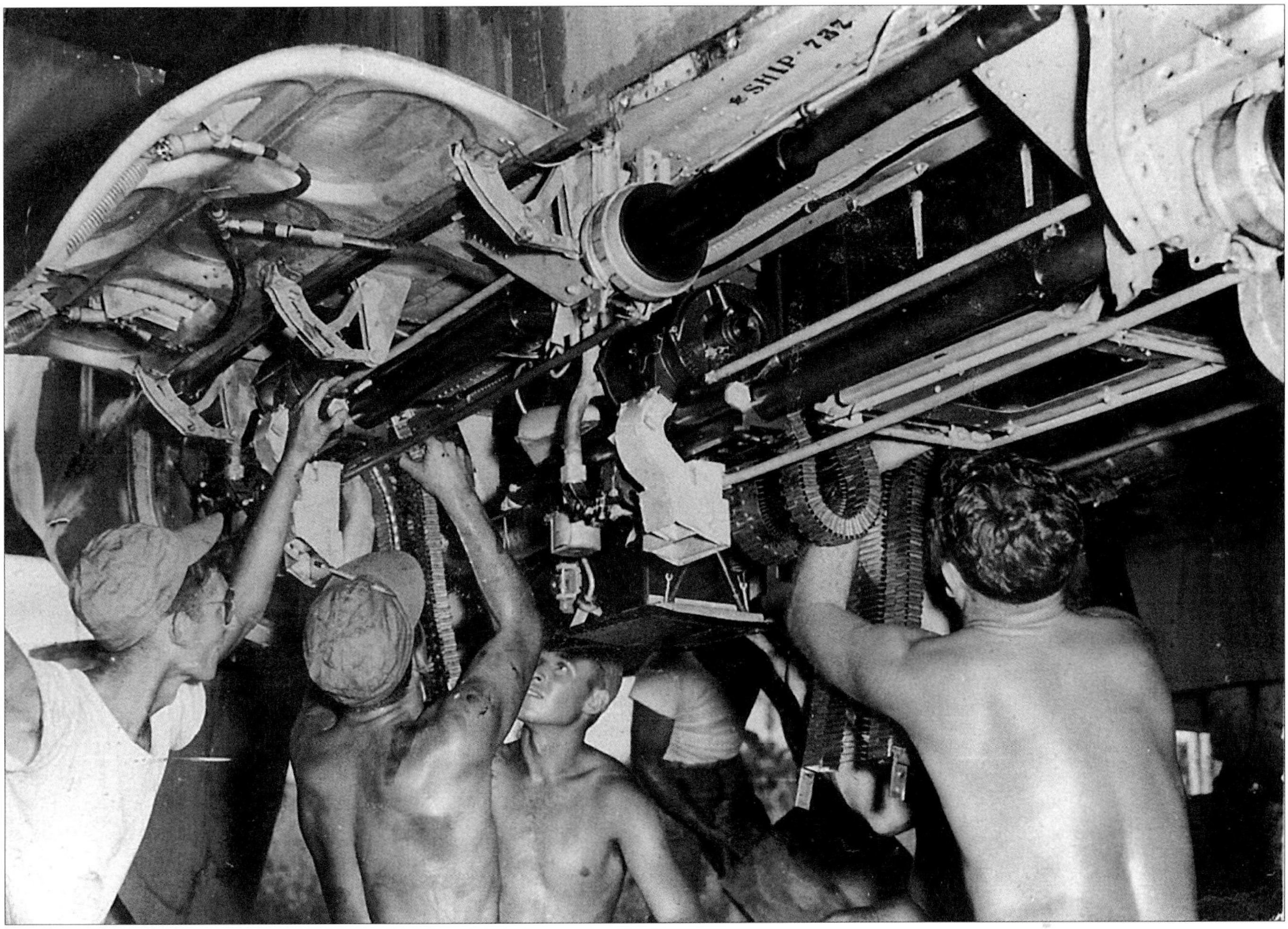

419th NFS personnel working on the 20mm cannon of a P-61 Black Widow. The P-61 mounted two separate gun systems, including four .50 caliber machine guns in an upper turret and four 20mm cannon in the belly. Later variants of the P-61 "A" model omitted the machine gun turret. It was re-added to later "B" models of the aircraft. *Courtesy of Leslee Lukas Mays.*

afloat until a rescue craft reached him. McCall had also flown as the gunner for Manke when his landing gear collapsed two nights earlier.

July 1944 was not a good month for Detachment A. The group lost its first air crew members only three days after beginning combat operations off New Guinea. Roney was the most decorated member of the unit, having transferred to the 419th NFS after a tour flying Curtiss P-40 Warhawks and Lockheed P-38 Lightnings during the Aleutian campaign, where he had received the Distinguished Flying Cross. Both lieutenants were very popular among the detachment personnel, and their deaths were bitter reminders of the dangers associated with flight operations. In three days of operations Detachment A's combat strength dropped from six to three P-61s. Lukas and his regular crew were up the next morning, providing dawn coverage for the Sansapor invasion fleet. While all grieved for Roney and Zint, the war continued.

Master Sergeant McConnell believed small particles of coral were the cause of the accident:

"Coral was very hard on the airplanes, and minute amounts of dust or particles would get into critical components of the engines. Maintenance checks of Roney's P-61 indicated that everything was OK before the plane took off. She rose, stalled out, and went in. We waved to a native nearby in a canoe and he rowed us out to the crash site."[3]

Sergeant Harold Young, a P-61 mechanic, verified the point. He noted:

"The coral was really harsh on the aircraft. In particular, coral dust was very abrasive. When we ran the engines on the ground, the propellers would pick up the dust and put little nicks into the blades."

Left to right: Crew Chief Staff Sergeant Thomas Mezo, Line Chief Master Sergeant Bill McConnell, and Engineering Officer First Lieutenant Michael Burlingham. McConnell believed the crash resulting in the deaths of First Lieutenants Roney and Zint was caused by coral dust in the engines. *Courtesy of Don Dessert.*

Coral dust also proved to be detrimental to the windshields and other kinds of glass or Plexiglas. Young remarked they had to keep the windshields covered anytime they were on the ground utilizing anything they could find.

Besides the fine coral dust, the mechanics had to contend with the many problems associated with working on aircraft in a tropical environment. Sea salt permeated the air, causing corrosion on electronic and electrical systems. The ground crews had to maintain a coating of oil on the engine propellers, which often picked up sea spray as the planes flew low over the ocean. Young noted that this task was not as easy as it sounded:

"We wanted a coating of oil on the propellers that would stay on them, but if you got it too heavy on the blades they would suck up all the dust in the area, and that coral dust was very abrasive."

High humidity and rain produced problems associated with mud, rust, and mold on the equipment.[4] First Lieutenant Bob Fowble, an R/O who joined the 419th NFS in May 1945, noted that the tropical conditions were especially detrimental to the sensitive radar equipment.[5]

Night operations were simply dangerous affairs for air crews and could result in non-combat related accidents at any time. Pilots flew combat weary aircraft that, despite the best efforts of the hard working ground crews, often experienced mechanical failures. The European Theater held priority for supplies, spare parts, and new equipment, including aircraft. Army Air Force personnel in the Pacific Theater continually struggled throughout most of World War II to secure their operational requirements. Mechanics frequently lacked sufficient spare parts to keep all of a squadron's aircraft airborne. Battle or accident damaged aircraft often became "hangar queens," a term for a plane relegated to duty as a source of spare parts to keep others in the sky. The supply process for the few available spare parts was slow and bureaucratic at best. The delay of a supply vessel of five days could mean five more days that combat aircraft were grounded for required maintenance or damage. Grounded aircraft meant fewer combat sorties over Japanese targets or airborne to protect American bases from Japanese fighters and bombers.

Battle damage to the outer surfaces of the aircraft was repaired by the squadron's sheet metal specialists. Staff Sergeant Elzie "Eddy" Ballard, an aircraft crew chief, commented that the sheet metal personnel could round out any holes in the airplane. They would then cut a piece of metal and rivet it over the hole in the aircraft. Ballard bragged that the 419th NFS had one of the best sheet metal teams, led by Technical Sergeant Thomas Camp, that he had seen in the Army Air Force.[6]

The Nighthawks and Black Widows had their own unique mechanical issues. Master Sergeant McConnell remembered that the P-61 was an easier aircraft to maintain than the Douglas P-70 Nighthawk. Supply problems combined with mechanical issues made the P-70 a difficult plane to maintain. "The radar unit on the P-70 would burn up the generators on the aircraft. Those generators were not easy to replace."[7] Sergeant Young noted that the P-70 was an older aircraft than the P-61. They were war weary, with older engines that tended to leak oil. According to Sergeant Young, one of the most common problems with the P-61 involved the exhaust system. "We had chronic problems with the exhausts of the Black Widows."[8]

The ground crews were a resourceful group of men who kept the planes flying despite the weather conditions and lack of spare parts. Sergeant Young quietly stated, "We tried to do the best we could under the conditions."[9] The pilots of the 419th NFS offered nothing but praise for these men who worked day and night to keep the planes mission capable. Dick Stewart rewarded his crew chief's hard work by bringing him a bottle of alcohol whenever returning from leave in Australia or New Zealand.[10] First Lieutenant Don Dessert echoed the comments:

"My crew chief was Staff Sergeant Leonard Keim. He was the tallest crew chief in the 419th NFS. He was so tall that he could hand a parachute to me in the cockpit while standing alongside the P-61 and without using a ladder. The maintenance on the P-61s under the adverse conditions was a credit to all our ground crews, and my plane, *Old Salty Dog II*, was never out of commission when needed. Before the arrival of thc P-61s the crew chief on my P-38, *Swoosh*, was Sergeant 'Wee Willy' Fellows. Again, *Swoosh* was never out of commission when she was needed. I can't say enough about these two men."[11]

Captain Al Lukas added another "second" to these comments:

"I have never seen a more dedicated group of individuals than our ground personnel. They put in longer hours repairing and maintaining our aircraft despite lacking many resources and spare parts. I provided many of them with rides in the P-61 as a way of saying 'thanks' for their efforts."[12]

The American airfield at Noemfoor following a July 1944 Japanese air raid. Note the B-24 Liberator bombers dispersed in the foreground of the photo and the fighters packed tightly in the center of the airfield. *Courtesy of Leslee Lukas Mays.*

Operations clerk Sergeant Jesse Gardner flew with Lukas:

"We flew straight up the coast, and then he did some buzz jobs along the shore line and followed the contours inland part of the way. It was really swell, and the first time I'd done any stunts in a plane."[13]

Sergeant Ralph Dunavant, a clerk with the intelligence section, benefited from the generosity of the pilots:

"The flying crews were always good to us. I tried to go up every place we were stationed. While at Sansapor I missed the last ferry to Middleburg Island. I called to explain the situation and one of the pilots flew over in the A-24 to pick me up. That's the kind of unit the 419th was. Everyone was willing to give you a hand."[14]

Japanese night fighters struck Noemfoor on a regular basis between the American assault on July 2, 1944, and the end of the month. Detachment A's personnel attempted to dig fox holes after arriving at the bivouac area but struck solid coral after only six inches. That same evening, a Japanese bomber came across the island and dropped its load before escaping unscathed. One of the bombs fell approximately 200 yards from the Detachment A area, where the men had little protection except to lie prone. Sergeant Charles "Buddy" Bailey related:

"I still believe if I had hugged the girl back home who sent me the 'Dear John' letter like I hugged mother earth that night, she would still be an old maid."[15]

First Lieutenant Manke conducted the first squadron P-61 pursuit of a Japanese night fighter on the night of July 30, 1944. Two Japanese raiders slipped into the area and bombed Noemfoor. The GCI VHF radio failed and Manke was not able to hear the GCI controller's instructions nor acquire the bogeys with the SCR 720 radar onboard the P-61. It proved to be a frustrating end for the first combat pursuit by the squadron.[16]

Captain Al Lukas made the first 13th Air Force P-61 Black Widow kill. Lukas never forgot that it was the hard work of the ground crews that kept the P-61s airborne in the tropical heat and rain. "I have never seen a more dedicated group of individuals than our ground personnel. They put in longer hours repairing and maintaining our aircraft despite lacking many resources and spare parts. I provided many of them with rides in the P-61 as a way of saying 'thanks' for their efforts." *Courtesy of Leslee Lukas Mays.*

The first four nights of August were quiet, with Detachment A providing night fighter support over Noemfoor as well as Sansapor without encountering any enemy aircraft. Japanese night fighters returned to Noemfoor three times in August beginning on the 5th, but this time the 419th NFS would be waiting to greet them on each occasion.

Captain Lukas, with First Lieutenant John Blankenship (R/O) and Staff Sergeant Glen "Frosty" DeForrest (gunner), bagged the 419th NFS' first kill of a Japanese intruder during the early hours of August 5, 1944. Lukas took off from Kamiri strip and commenced a routine night patrol 30-40 miles south of Noemfoor. Nearly two hours later the trio received a warning from First Lieutenant Dave Guy, a GCI controller attached to the unit. Guy informed the crew that his station had a radar fix on an unidentified bogey heading toward Noemfoor. Guy directed Lukas to maneuver his P-61 to a vector of 120 degrees in order to intercept the unknown contact. The pilot complied, and was soon informed to alter his approach to a vector of 90 degrees and to begin losing altitude. At 2:00 AM Blankenship enthusiastically reported a contact on his SCR 720 airborne radar set. Although limited in range to a few miles, the radar unit provided the crew this night with the necessary edge to find and make visual contact with the unidentified aircraft being tracked by ground based radar.

The bogey was eight miles away and approaching Noemfoor. Lukas boosted the P-61 to full throttle to catch the bogey before it dropped any ordnance on the American airfield. The unknown aircraft initiated a sharp turn at the southeast edge of Noemfoor as the P-61 stealthily approached from behind. Lukas eased back on the throttle at the direction of his radar observer and peered into the night sky for a visual contact. Blankenship reported the bogey should be 400 feet ahead and slightly starboard of its pursuer. Lukas throttled back again and spotted a dark object silhouetted by the moon. The crew easily identified the bogey as a two engine Japanese Sally bomber. Although Sally bombers were fairly obsolete for most day missions by 1944, the Japanese still utilized the flying workhorse for night bombing missions of Allied

Captain Al Lukas (right) standing in front of 419th NFS P-61Black Widow 42-5505 #310 *Midnite Majorie: Luke's London Lovely*, in which he downed a Japanese Mitsubishi Ki-21 Sally bomber on August 5, 1944. This photo provides a good look at the nose art Lukas placed on the aircraft for his English wife, Marjorie Kingsley. Most 419th NFS aircraft did not carry any personal nose art or names other than the official three digits added upon assembly in Guadalcanal. Even these numbers were moved to the rear of the aircraft by the time the squadron arrived in the Philippines. *Courtesy of Leslee Lukas Mays.*

A Mitsubishi Ki-21 Sally bomber. Captain Al Lukas (pilot), First Lieutenant John Blankenship (radar observer), and Staff Sergeant Glen DeForrest (gunner) became the first 13th Air Force crew to down an enemy aircraft while flying the P-61 Black Widow on August 5, 1944. Both Lukas and DeForrest fired upon the Japanese Sally bomber. The pilot reported, "The Sally's port engine by this time was licking flames. Both the gunner and I then fired a one and a half second burst with the four 20 mm cannon and the four .50 caliber turret guns, blanketing the Sally...The gunner and I again opened fire with a final one second coup de grace that distinguished his empennage and left both engines burning steadily." *Courtesy of Leslee Lukas Mays.*

airfields. Lukas maintained his altitude and airspeed for nearly thirty seconds as the crew surveyed their target, apparently unnoticed by the Japanese crew.

Lukas eased back on the throttle for the third time, maneuvered behind the Japanese bomber, and informed the gunner to fire the four .50 caliber machine guns located in the turret. The first burst struck the Sally's left engine and left wing. The Japanese crew decreased altitude and sharply turned their plane to port in an attempt to escape the unseen attacker. Lukas pursued the Sally and moved to place the Japanese plane 400 feet in front and slightly above him. The left engine of the Sally could be seen trailing a yellow flame from the successful .50 caliber strikes. Lukas reported:

"The Sally's port engine by this time was licking flames. Both the gunner and I then fired a one and a half second burst with the four 20 mm cannon and the four .50 caliber turret guns, blanketing the Sally."

The Sally turned to starboard and dove toward the ocean. Lukas followed the maneuver and delivered a two second burst from both sets of guns that left the enemy aircraft enveloped in flames:

"The gunner and I again opened fire with a final one second coup de grace that distinguished his empennage and left both engines burning steadily."

At 7,500 feet, Lukas ended his dive and watched the Sally hit the water approximately 20 miles southeast of Noemfoor Island. The crew had fired 160 rounds of 20 mm cannon and 325 rounds of .50 caliber ammunition during the kill.

Although the kill looked easy in a retrospective examination, the GCI/P-61 team demonstrated incredible coordination in locating, tracking, and downing the Sally bomber. The Black Widow crew continued its patrol for another 40 minutes in case a second Japanese night fighter lurked south of Noemfoor. The crew then landed with the satisfaction of having become the first 13th Air Force crew to shoot down a Japanese plane with a P-61 Black Widow. Lukas offered praise for Guy when he noted, "Perfect GCI vectoring to position the night fighter is not an easy feat." The Black Widows of the 419th NFS had woven their web across the northern coast of New Guinea and dared the Japanese to test their airspace.[17]

Lukas landed to a roaring reception from his squadron mates on night duty that evening. First Lieutenant Bob Oates, with R/O Flight Officer Pete Fulghum and gunner Steve Lackey, took up a P-61 to continue the patrol of the area after the arrival of Lukas. At 3:30 AM the GCI controller reported another incoming bogey. Oates altered his heading to intercept the unidentified aircraft. Fulghum established his own radar fix and

First Lieutenant John Blankenship, the Radar Observer (R/O) for Captain Al Lukas during the downing of a Japanese Mitsubishi Ki-21 Sally bomber near Noemfoor in August 1944. Blankenship flew as the R/O for 419th NFS commander Captain John McCloskey during the November 1943 P-70 Nighthawk training accident at Guadalcanal. First Lieutenants Don Dessert and Dick Stewart, along with Corporal Theodore Knutson, swam out to the P-70 wreck to save the crew, who were trapped in the aircraft. Fuel poured into the surrounding water as Dessert and Stewart pulled Blankenship and First Lieutenant George Budway from the wreck. However, they were not able to reach McCloskey, who died within the inverted plane. *Courtesy of Don Dessert.*

First Lieutenant Bob Oates standing in front of 419th NFS P-61Black Widow 41-5508 #313. Oates lost a bogey near Noemfoor when the Black Widow's SCR-720 radar system malfunctioned. While on Morotai, Oates was scrambled in a P-61 by the GCI controller to intercept an incoming bogey.The enemy aircraft arrived before Oates could get airborne to engage him. The airfield on Morotai consisted of two parallel landing strips. As Oates was taking off on Wama Airstrip, a Japanese fighter was flying parallel to him and strafing Pitoe Airstrip. If the Japanese intruder had chosen Wama Airstrip he would have caught Oates during the vulnerable time when engaged in a takeoff. *Courtesy of Leslee Lukas Mays.*

directed Oates from this point. As the pilot neared the enemy aircraft, the P-61 crew flew into the Noemfoor AA circle. The local AA battery initiated a defensive barrage against the unknown interloper and at the same time placed the P-61 in potential peril. Oates broke off contact with the other aircraft and orbited in the area for it to be destroyed or emerge from the AA circle. At this time the P-61's SCR720 radar set malfunctioned. Fulghum remembered crying out "Aw shit!" into the intercom system and hearing a similar reply from a frustrated Oates. The R/O could not acquire another local radar fix as the bogey departed the AA circle and withdrew from the area. Oates landed, and Lukas commenced his second early morning patrol. The Japanese intruders did not return. Dawn on the morning of August 5, 1944, ended a night of jubilation for Lukas and frustration for Oates. This would not be the last time that an equipment failure robbed a 419th NFS pilot of a possible aerial victory against a Japanese night fighter.

Sergeant Edwin "Eddie" Barkhouse remembered Lukas as a man "with a big heart." After shooting down the Sally bomber, Lukas assessed each of the aircrew members to provide a contribution of alcohol. He then asked for a gift of medicinal alcohol from the medical personnel. Lukas mixed the various types of alcohol in a 20 gallon cooking pot with fruit juices and threw a party for the entire detachment.[18]

Two nights after Lukas scored the first squadron kill, First Lieutenant Manke scrambled from Noemfoor in pursuit of an unidentified bogey north of the island. A problem with his communications equipment nearly brought down his P-61 to friendly fire that evening. Manke climbed toward the south after taking off from Noemfoor. However, his communications reception on the VHF radio set was very weak. The pilot could not understand the instructions being relayed by the GCI controller. He heard an occasional clear vector over the set and followed those instructions. The vectors were directing the crew to turn back toward Noemfoor. An unidentified aircraft lurked somewhere in the dark and the P-61 could not find it. The R/O, Flight Officer Sandy Bauman, could not get a fix on his radar, and the malfunctioning radio set prevented clear communications with the GCI controller on Noemfoor. The island's anti-aircraft (AA) battery commenced firing as Manke neared Noemfoor, and the pilot initially believed the AA gunners were aiming their ordnance at him. However, Bauman finally acquired a radar fix on the bogey. The unidentified aircraft was 2,000 feet above them and only one mile ahead. Manke continued the pursuit although he had entered the AA circle ring around the island. The pilot and his gunner, Sergeant Leroy Piatt, made visual contact with the enemy aircraft as anti-aircraft shells began to burst around the P-61. Manke made a hard turn to port to avoid being brought down by friendly fire. The crew lost sight of the other aircraft after the hard maneuver and returned to their former compass heading. Bauman regained a faint fix on the bogey, and the chase continued for a minute until the Japanese aircraft dropped off the radar scope. The

Left to right: Flight Officer Malcolm Holloway and Second Lieutenant Phil Rand. Holloway flew as Rand's Radar Observer (R/O) on an August 1944 mission when the pair exchanged gunfire with a Japanese Mitsubishi Ki-46 Dinah. Before Rand could maneuver the P-61 into a firing position, the Dinah opened up on him with its twin dorsal 7.7 mm machine guns. Tracers raced past the Black Widow's port engine. Rand believed the rounds struck his aircraft and initiated an evasive maneuver to slip out of the stream of enemy bullets. Rand called for Holloway to re-establish radar contact with the Dinah. The R/O reported a new radar fix and the pilot continued the pursuit. After a brief chase the Dinah managed to evade detection by Holloways' radar set. Rand turned back toward Noemfoor and landed at 4:30 AM. An inspection of the aircraft did not reveal any damage from the encounter with the Dinah. *Courtesy of Don Dessert.*

R/O acquired a third fix and Manke renewed the pursuit. As he approached the target, Manke fretted and made the tough decision to halt the chase. The P-61 had not been refueled after its early evening patrol and Manke did not want to risk running short of fuel and having to ditch in the ocean. The entire scramble lasted only 45 minutes and proved to be a frustrating example of the requirement for operable communications equipment to ensure a night intercept of Japanese intruders.

The next night, August 8, 1944, the unit tapped Second Lieutenant Phil Rand to fly the early morning P-61 patrol. Rand launched his Black Widow at 2:30 AM and patrolled south of Noemfoor. Staff Sergeant DeForrest, the regular gunner for Lukas, rode behind Rand, and Flight Officer Malcolm "Mac" Holloway sat in the R/O compartment. The patrol seemed routine, however, Rand would become the first 419th NFS pilot to face defensive fire from a Japanese plane before the end of the morning. Approximately 90 minutes after beginning his patrol, the GCI controller radioed a bogey contact south of the island. Rand turned his aircraft and commenced a chase based on the vector provided by the GCI controller. The P-61 crew acquired a visual of the other aircraft as it neared the southeast coast of Noemfoor. The bogey flew only 200 feet below and approximately 600 feet off their port wing. The P-61 crew confirmed the bogey as a Japanese Mitsubishi Ki-46 Dinah. Before Rand could maneuver the P-61 into a firing position, the Dinah opened up on him with its twin dorsal 7.7 mm machine guns. Tracers raced past the Black Widow's port engine. Rand believed the rounds struck his aircraft and initiated an evasive maneuver to slip out of the stream of enemy bullets. A check of the cockpit gauges indicated a drop in temperature within the port cylinder head. The Dinah dove in an attempt to extract itself from the ambush. The P-61 seemed to still be in flying condition, so Rand called for Holloway to re-establish radar contact with the Dinah. The R/O reported a new radar fix and the pilot continued the pursuit. The Dinah dove again from 15,000 to approximately 9,000 feet, managing to evade detection by Holloway's radar set. Rand turned back toward Noemfoor and landed at 4:30 AM. An inspection of the aircraft did not reveal any damage from the encounter with the Dinah.

Japanese night fighters returned for a third night during the month when the GCI controller acquired a radar fix on an unidentified bogey around 4:30 AM on August 9, 1944. The GCI team directed Captain Lukas, the orbiting P-61 pilot, to alter his course for an intercept of the unknown aircraft flying between 30 and 40 miles southeast of Noemfoor. The bogey did not get any closer to the island, and the P-61 crew never made a local acquisition of it on their radar set. This proved to be a frustrating scenario for the pilot, who itched to score his second kill in one week. Japanese night raiders did not attempt another intruder mission against Noemfoor until September.

A tragic accident occurred on August 14, 1944. First Lieutenant Don Gardner sat in the cockpit of his P-61 at Kamiri Airstrip awaiting the return of a P-61 flown by First Lieutenant George White. A heavy rain squall

First Lieutenant Don Gardner died in an accident on Noemfoor. Gardner sat in the cockpit of his P-61 at Kamiri Airstrip awaiting the return of a P-61 flown by First Lieutenant George White. A heavy rain squall passed across the island as White landed his Black Widow, making the coral runway very slippery. White's plane began to skid uncontrollably at the end of the runway, and the starboard wing of the P-61 sheared the nose and cockpit of Gardner's plane, killing the pilot instantly where he sat. Captain Al Lukas displayed his leadership qualities and immediately arranged a military funeral for Gardner to help bring closure to the incident. Lukas wrote in his flight log, "What a lousy break for one of the finest lads in this unit." *Courtesy of Leslee Lukas Mays.*

First Lieutenant George White piloted the P-61 Black Widow that entered an uncontrollable skid after landing on the wet coral of Noemfoor. His plane sheared the cockpit and nose of a second P-61 in which First Lieutenant Don Gardner sat. The accident and death of Garner devastated the morale of the unit, and especially White, who sat helplessly in the P-61, unable to control the plane as it slid across the wet coral. Air and ground crews noted that there was nothing that White could have done to prevent his plane from striking Gardner's P-61. White was never quite the same after the accident. *Courtesy of Don Dessert.*

passed across the island as White landed his Black Widow. The rain made the coral runway very slippery, and White's plane began to skid uncontrollably at the end of the runway. The starboard wing of the P-61 sheared the nose and cockpit of Gardner's plane, killing the pilot instantly where he sat. The accident devastated the morale of the unit and especially White, who sat helplessly in the P-61, unable to control the plane as it slid across the wet coral. Captain Lukas displayed his leadership ability and immediately arranged a military funeral for Gardner to help bring closure to the incident. Lukas wrote in his flight log, "What a lousy break for one of the finest lads in this unit." Gardner's plane was considered unsalvageable and was relegated as a "hangar queen" for parts.

During late August, Detachment A moved its flight line from Kamiri Strip to Kornasoren Strip, and by the end of the month the ground and air crews agreed they had developed the best flight line at the Strip. The men constructed a combined operations/intelligence/engineering/pilots' alert hut, an armament shop, a sheet metal shop, an electrical shop, a technical supply hut, a motor pool, a communications shop, a radar shop, and a DF Homing Station. Kornasoren was a newly opened airstrip, permitting Detachment A to actually plan the layout of their area. Captain Warren Slater, the Squadron Intelligence Officer, noted:

"The flight line is ideally located on an oblique taxiway 500 feet off the end of the strip. Thus, the P-61s are parked in a place of safety from landing aircraft and in a most favorable position to facilitate expeditious take-offs on night scrambles."

The new aircraft parking arrangement at Kornasoren Airstrip prevented a repeat of the landing accident that occurred at Kamiri Airstrip on 14 August. Detachment A flew a total of 102 sorties during 75 missions in August with only five P-61 Black Widows. The recreational highlight of the month was the arrival of a USO show. Although Noemfoor was the most forward American-held area in New Guinea during the previous month, Bob Hope brought his USO tour to the island. He performed with Jerry Calona, Frances Langford, and Patsy Thomas. It's easy to understand why Hope would be granted "honorary military veteran" status by the United States government over sixty years later.[19]

The coral composition of the island and its low altitude above sea level made the construction of latrines a challenge. First Lieutenant Robert "Bob" O'Brien, the Ordnance Officer, explained:

"On Noemfoor, the water table was about four feet below ground level, so digging or blasting latrine holes was not an option. Someone came up with a great idea to cut the tops and bottoms out of 55 gallon drums and then stand them up and building a platform of coral around them. We added steps up to them and a roof over them. To clean them, diesel fuel was dumped in and burned every day. We also built an outdoor movie theater using fallen coconut logs for seats. Unfortunately, it turned out that you could see the movies really well from the raised seats of the latrine and it was hard to get users to do their business and leave."[20]

Sergeant Charles "Buddy" Bailey added another detail about latrine construction:

"One morning Private Ernest Stumpo was on squadron detail, which included burning out the officers' latrine. Stumpo poured the mixture of fuel and oil in the hold and reached into his pocket for a match. Not finding any, he went back to his tent for some matches. In the meantime, one of the new officers went up to the latrine, lit his cigarette, and threw the match into the hole. I heard a sound and stepped out just in time to see the officer bail out with his pants around his knees. He wasn't burned, but his pride sure was singed."[21]

Detachment A personnel continued the old military tradition of trading to acquire items they needed or simply wanted. Arranging elaborate trades, often consisting of odd items, lifted the morale of the men and gave them something to do when off duty. First Lieutenant Bob O'Brien, the Squadron ordnance officer, remembered members of the Detachment trading various items with other units in the area. On one occasion, someone in the Detachment arranged a trade where he acquired an ice cream machine to the delight of others. Despite the joy of hard to get items like ice cream, alcoholic beverages remained "must acquire" drinks simply because of their lack of availability. O'Brien noted that at least on one occasion someone in the Detachment traded with the Navy for alcohol utilized in torpedoes which was, in turn, utilized to produce "home brew." When pilots returned from their quarterly leaves to Australia they would bring back any type of alcoholic beverage available at the time of their visit. O'Brien remarked:

The Radar Observer sat in his own compartment in the rear of the P-61 Black Widow. He faced forward where his radar equipment was located. *Courtesy of Leslee Lukas Mays.*

Preparing a 419th NFS P-61 Black Widow for a mission. The squadron later darkened the radar dish covers on the nose of their aircraft. *Courtesy of Don Dessert.*

Second Lieutenant James Schroth (far left) sitting on the wing of a P-38 Lightning. Schroth, along with Radar Observer Flight Officer Fred James and gunner Staff Sergeant Glen DeForrest, downed a Japanese Mitsubishi Ki-46 Dinah in September 1944 while flying from Noemfoor. The Dinah initiated a dive and Schroth followed closely behind the Japanese plane. "I nosed the P-61 down and followed the enemy while my gunner fired a long five second burst with the four 50s in the turret, hitting the Dinah's port engine, which began burning. The Dinah made a gentle port turn and a 20 mm dorsal gun fired at us. The tracers went by low, to starboard, and fairly close. Both the gunner and I replied with the four 50s and the four 20s from astern and slightly below. We covered the Dinah pretty well with direct hits. The Dinah went spiraling down in flames from 11,000 feet until it disappeared below the overcast, which was 2,000 feet below us." *Courtesy of Leslee Lukas Mays.*

"We made 'Purple Passion' in large pots from the kitchen by combining grape juice with whatever drinkable stuff the pilots and air crew brought back from Australia. This was serious drinking, and hangovers could last for days."[22]

Japanese intruders returned to Noemfoor with the full moon on the first day of September 1944. The GCI controller scrambled First Lieutenant James Schroth and his crew, Flight Officer Fred James (R/O) and Staff Sergeant DeForrest (gunner), at 9:30 PM. At 9:45 PM the controller vectored the P-61 on a heading of 110 degrees and then changed that to 180 degrees within minutes. After flying approximately twenty miles, Schroth received a radio call to alter his course to 230 degrees. Ten miles later the GCI controller directed a change to 250 degrees. Following another ten miles, James made contact with the bogey at a range of five miles on the SCR 720 radar set. The bogey was flying between 4,000 and 5,000 feet in altitude and was now 25 miles south of Noemfoor Island.

As Schroth dived to intercept the unidentified aircraft James lost contact on the radar. When the P-61 reached the correct altitude, the GCI controller provided a new heading until James, again, established his own radar contact. The R/O directed Schroth to a position within 500 feet of the bogey. At this point DeForrest established visual contact with the prey. Schroth eased the P-61 forward until he was within 400 feet, and then he too could see the enemy aircraft they were chasing. Schroth identified the Japanese aircraft as a Dinah. The pilot ordered DeForrest to open fire with the .50 caliber machine guns in the turret. Schroth noted, "The gunner fired a 2-3 second burst and saw strikes near the port wing root." The Dinah initiated a dive and Schroth followed closely behind the Japanese plane:

"I nosed the P-61 down and followed the enemy while my gunner fired a long five second burst with the four 50s in the turret, hitting the Dinah's port engine, which began burning. The Dinah made a gentle port turn

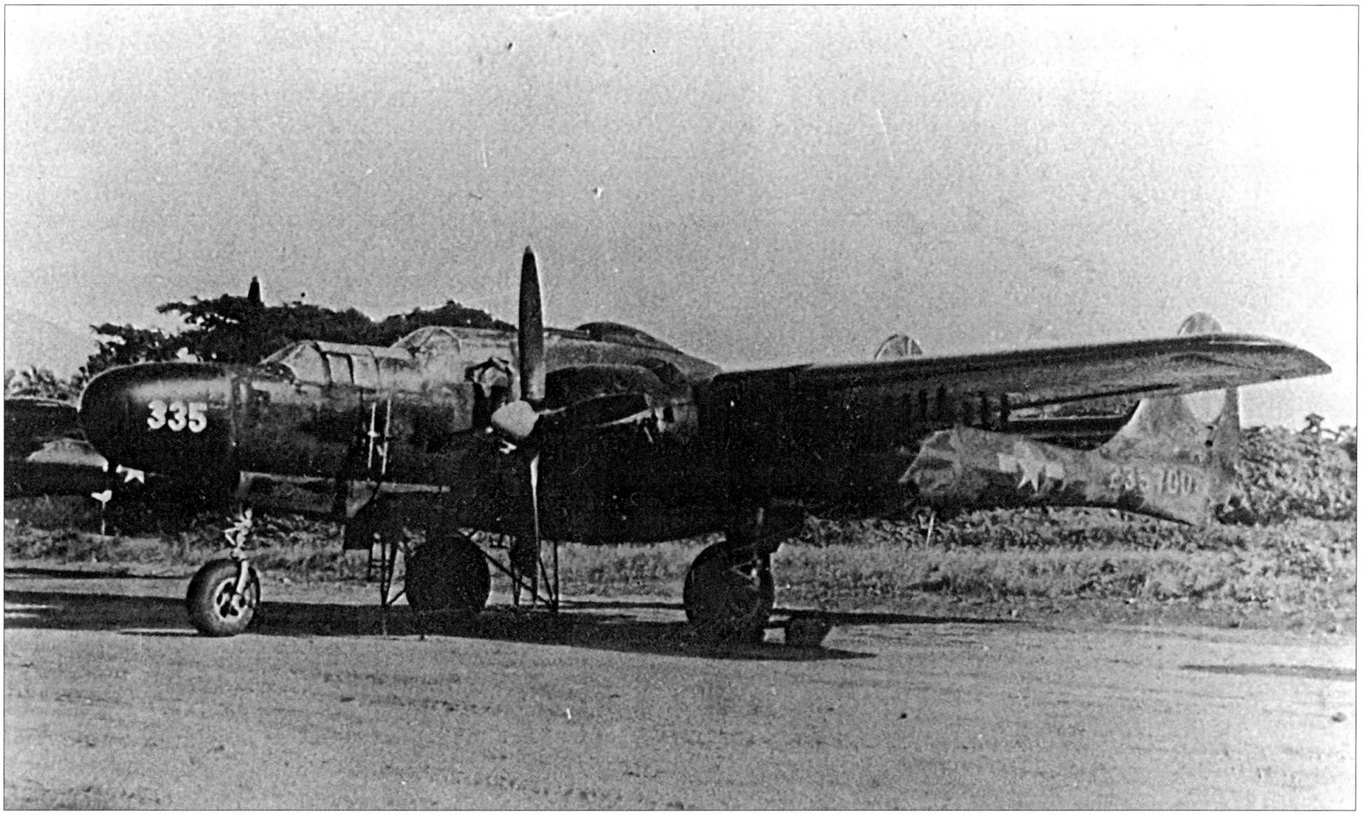

419th NFS P-61 Black Widow 42-39700 #335. The pilot and gunner of a Black Widow entered their compartment through an access panel behind the nose wheel of the plane. A panel with a ladder on its inside dropped down, allowing the two crew members to climb up and into the aircraft behind the pilot's seat. The radar observer had his own access panel and ladder that dropped down from the floor of his compartment at the rear of the aircraft. *Courtesy of Don Dessert.*

and a 20 mm dorsal gun fired at us. The tracers went by low, to starboard, and fairly close. Both the gunner and I replied with the four 50s [.50 caliber machine guns] and the four 20s [20 mm cannon] from astern and slightly below. We covered the Dinah pretty well with direct hits. The Dinah went spiraling down in flames from 11,000 feet until it disappeared below the overcast, which was 2,000 feet below us."

Schroth reported a "kill" to the GCI controller and asked for his current position. The controller replied that the P-61 crew was near the Japanese Moemi Airdrome and provided a vector for home. The Black Widow returned safely and landed at 11:30 PM. The P-61 expended 80 rounds of 20 mm cannon and 790 rounds of .50 caliber ammunition in making the kill. The crew gave special credit to the GCI controller, First Lieutenant Dave Guy, for assistance in downing the Dinah. Guy was officially attached to the 419th NFS, and his carefully calculated and continuous vector changes ensured that Schroth was able to intercept the Japanese aircraft at a location where James could take over with his SCR 720 system. Schroth added:

"The GCI controller, First Lieutenant Dave Guy...deserves special commendation for the extremely efficient manner in which he vectored us to the target and then after we lost the initial contact, again vectored us to the target...Successful night fighting depends on excellent team work by the GCI controller, radar observer, gunner, and pilot."

Counting his work with other units, this was the fourth time Guy had helped direct a night fighter team to a kill, and he would later earn a Bronze Star medal for his GCI work. One should also not forget that DeForrest served as Captain Lukas' gunner the night they bagged a Sally. Thus, he now had two confirmed kills as the gunner for two different pilots.[23]

Another "first" for the 419th NFS occurred in the early hours of September 9, 1944. The GCI controller alerted the squadron that he had acquired an unknown aircraft on the radar set at 2:20 AM. Detachment A immediately scrambled two P-61 teams. First Lieutenant George H. White and First Lieutenant Manke launched their Black Widows in pursuit of the bogey. The GCI controller directed White and Manke

419th NFS P-61 Black Widow 43-8238 #332 *Vivacious Vivian.* This P-61 was one of the very few 419th NFS P-61s that sported nose art. *Courtesy of Don Dessert.*

within three miles of the unidentified aircraft, but neither P-61 crew could locate it on their radar sets. The bogey did not approach any closer to Noemfoor and the pair of P-61s returned to their roost.

Scrambling a pair of P-61s was an infrequent event in the 419th NFS due to several factors, including the lack of available aircraft and the need to allow crews to rest in order to remain sharp when flying these hazardous night missions. At the end of July Detachment A could count on only three serviceable P-61s, but a month later this situation had changed. The Detachment now operated five serviceable P-61s. On this particular night two P-61 crews were available for the scramble. Working as a team permitted the crews to separate and coordinate their own radar sets in cooperation with the GCI controller on Noemfoor. The first P-61 to make a local radar fix would pursue the bogey, with the second P-61 providing support. If the SCR 720 radar malfunctioned the second plane could immediately continue the chase. In addition, if the pursuing P-61 lost its radar fix before making a visual sighting of the bogey, the second Black Widow might be able to maintain or make its own radar acquisition and assume the lead. Throughout most of September Detachment A maintained two P-61s on alert status each night.

On September 16, 1944, First Lieutenant Robert Williamson scrambled with his crew to intercept an unidentified aircraft reported by the GCI controller at 3:30 AM. The bogey passed south of Noemfoor and appeared briefly on the P-61's SCR 720 radar. Williamson's crew was unable to make a second fix and returned to Noemfoor. After this brief chase the remainder of September proved to be a relatively quiet period for Detachment A.

After the rough months of July and August, Detachment A seemed to settle into an acceptable routine on Noemfoor in September. The morale of the personnel increased for many reasons. The detachment did not suffer any aircraft accidents, and all of their P-61s remained in serviceable condition throughout the month, flying 96 sorties. Japanese night raiders did not drop any bombs on Noemfoor in September, and this is partially attributed to the presence of Detachment A and its Black Widows.

Left to right: Flight Officer Tony Grotzinger, First Lieutenant Herman Manke, and Second Lieutenant Bob Winslow. Note how the rear of the Radar Observer's compartment in the P-61 drops down to double as a ladder. Winslow was the second pilot who buzzed the high school football game with First Lieutenant Don Dessert in the Orlando area in 1943. In a second caper with Dessert, the duo joined Flight Officer Sandy Bauman for an excursion into a destroyed ammunition dump that had been declared off limits on Guadalcanal. Winslow dropped a hand grenade detonator that sent small slivers of metal flying into Dessert's legs, necessitating a confidential visit to "Doc" Felder for removal. Grotzinger flew as the Radar Observer for Winslow during a December 1944 mission out of Morotai when the pair received credit for a "probable" kill of a Japanese Mitsubishi G4M Betty bomber. During engagement the Betty's gunners returned fire as the P-61 maneuvered for a kill. A post-operation inspection of the P-61 revealed that the Japanese gunner had been fairly accurate. Winslow added, "I did not realize my plane was damaged until an examination after landing, which disclosed that the main spar and spoiler on the left wing were severely damaged by six 7.7 mm machinegun projectiles." *Courtesy of Garry Pape via Jeff Kolln.*

Obviously, the Japanese knew that American night fighters lurked in the skies around Noemfoor. The deterrent factor provided by Detachment A was a morale booster for the entire island. Food improved on Noemfoor, and First Lieutenant John Guzey, the Assistant Intelligence officer, bragged of an increase in fresh meat and eggs, as well as freshly baked bread in the mess hall. An issue of 12 bottles of beer per man "brought much joy to the hearts of all personnel." The receipt of the first combat leaves since June could be seen as the most important morale booster for the aircrews, who were now authorized 10 days in Sydney, Australia, plus travel time. Lukas departed on September 9, 1944, with the first group, while First Lieutenant Laffey filled in as the acting detachment commander.[24]

Operationally, the first few days of October were quiet. Japanese intruders returned beginning on October 8, 1944. Multiple enemy aircraft slipped past the radar net and dropped a total of three bombs on Noemfoor, wounding three men. However, the local AA battery managed to bring down two of the Japanese bombers over the island. The next night, between four and eight Japanese bombers returned to Noemfoor and bombed Kornasoren Airstrip, wounding three more men. First Lieutenant Manke was on patrol that night and the GCI

Left to right: First Lieutenant Bob Oates, First Lieutenant Aaron Jacobs, Flight Officer Harold Ozmun, Flight Officer John Radzieta, Flight Officer Lane K Thompson, and Flight Officer Roger Egan on leave at a New Zealand race track. Ozmun flew as the Radar Observer (R/O) for Captain Howard Daniel when their P-61 Black Widow disappeared on October 18, 1944, during an intruder mission bound for Waigeo Island. Thompson served as the R/O for First Lieutenant Don Dessert when they, along with gunner Staff Sergeant Frank McCormack, downed a Japanese Mitsubishi Ki-26 Dinah on October 2, 1944. *Courtesy of John Radzieta.*

controller vectored him to one of the incoming bogeys. The R/O made contact with the approaching aircraft at a range of two miles and directed Manke toward an interception course. However, the R/O quickly called forward to his pilot to warn that a second bogey was now on their tail. Manke temporarily abandoned the chase and maneuvered out of the way of the second plane. The R/O established radar contact with a new bogey and the crew initiated a pursuit. The chase ended as the enemy bomber entered the AA circle. The GCI controller then directed the P-61 to another incoming bogey. Manke flew an interception course until the local Fighter Sector ordered all friendly aircraft out of the area. The Noemfoor AA battery accounted for two additional Japanese bombers that night.

Detachment A flew 76 sorties during the month of October 1944. Nine of these sorties were flown as protective cover over B-24s that were taking off for long range bombing missions. At take off the bombers were slow, full of fuel and bombs, and easy prey for Japanese night intruders. Placing P-61s into orbit over the island as the bombers departed added a measure of safety to their mission.[25]

November brought orders for Detachment A to move forward to Morotai Island, halfway between New Guinea and the Philippines. As a result of the move the detachment flew only 26 sorties over Noemfoor in November. Everyone in the unit seemed excited about the prospect of transferring to a more active area where the Black Widows could find better hunting.

5

MIDDLEBURG

"The Black Widow is a formidable creature. It lurks in the dark... attacks unseen...cuts down its prey with poison 15 times stronger than rattlesnake venom."

Northup advertisement for the P-61 Black Widow

Allied forces were successfully securing the islands of Biak and Noemfoor at the end of July 1944. The next logical step in Allied plans following assaults on these islands involved establishing a foothold on the Vogelkop Peninsula of Western New Guinea. The Japanese maintained major bases on the peninsula, including Sorong, in the far west of the land mass, and Manokwari, which lay just 70 miles due west from Noemfoor. General Douglas MacArthur envisioned moving Allied forces as far west as possible and placing them in a position to directly support an invasion of the southern Philippines. In the process, the Allies would leapfrog past one or more of the major Japanese bases and simply allow them to waste away due to the inevitable cuts in their supply lines. Logically Manokwari would be bypassed, but plans to advance directly to Waigeo Island and cut off Sorong could not be completed due to the lack of a suitable area for airfields. The Allies settled on an area incorporating Sansapor and Mar, located approximately 70 miles east of Sorong. Plans called for the construction of a bomber field at Mar and a fighter field on the island of Middleburg, located approximately three miles off the New Guinea mainland. Five squadrons of medium bombers would be based at Mar and five squadrons of fighters on Middleburg. To provide night protection against Japanese intruders, Allied commanders opted to also dispatch a half squadron of night fighters on Middleburg.

Headquarters Detachment of the 419th NFS received the nod to make the move from Los Negros to Middleburg in early August 1944. Detachment A of the 419th NFS, on Noemfoor, provided nocturnal coverage of the landings at Sansapor, Mar, and Middleburg. Elements of the Army's 6th Infantry Division came ashore at Mar and Middleburg on 30 July and then Sansapor the next day. The Division encountered no Japanese resistance, but did capture considerable equipment, indicating a hasty retreat from the area. Airfield construction began quickly, with the fighter strip on Middleburg completed on 17 August and the bomber strip at Mar ready on 3 September.

The personnel of the 419th NFS Headquarters Detachment arrived off Middleburg in August 1944 on board the *USS John Lykes*. The Navy completed the offload of the squadron equipment on 10 August while the unit headquarters transferred to Sansapor and awaited further instructions. Detachment B, at Los Negros, in the Admiralty Islands, loaded its first contingent onto thirteen Douglas C-47 Skytrain transport aircraft on 18 August. They arrived at Middleburg on the same day, one day after the completion of the fighter strip. Aircrews also ferried four Northrop P-61 Black Widows and the detachment Douglas A-24 Banshee from Los Negros that day. The rest of the detachment personnel and equipment flew into Middleburg during the next two days. The Headquarters Detachment joined Detachment

Left to right in flight gear: Captain Al Lukas and First Lieutenant Don Dessert standing on the 419th NFS' Douglas A-24 Banshee. The A-24 was the Army version of the Navy's SBD dive bomber. The 419th NFS utilized this A-24 for instrument training and short hops between airfields. When one detachment of the 419th NFS transferred to Middleburg Island, the A-24 emerged as the critical element in a very important mission—moving the detachment beer supply. First Lieutenant Dick Stewart and Flight Officer James Allen packed as many cases of beer as possible in all storage locations on the plane. While weighted down in beer, Stewart and Allen encountered several Japanese fighters. Allen remembered, "Dick said, 'There's a flight of Jap fighters at 10 O'clock.' We both sat very still and they didn't see us. To this day, I don't know how they missed us." *Courtesy of Leslee Lukas Mays.*

B on 21 August, and the two organizations merged to become the Headquarters Detachment, 419th NFS with Detachment A on Noemfoor.

Everything belonging to the detachment found itself transported to Middleburg—even the beer rations. Flight Officer James "Jim" Allen admitted to transporting a plane load of beer in the unit A-24 with First Lieutenant Richard "Dick" Stewart piloting the aircraft:

"The outfit was moving from The Admiralty Islands to Middleburg, and Dick and I volunteered to take the A-24. We checked with the Navy before we left, and evidently they had just received a load of beer, and said we could have several cases. We took out everything that would come out of the compartments in the rear of the plane and put in the beer. We took the rear stick out, and I sat with my feet very high on two or three cases. You can imagine what that weight in the rear did for the plane. We got off OK, but when we started to land on some base to refuel it got very exciting—the nose jerked up and then down as Dick tried to get it under control at the slow landing speed. Somehow he managed to get it on the ground. A refueling truck came out, and as the operator was putting fuel in the plane I opened the storage compartments to check the beer. The Sergeant asked, 'What is that?' I said 'Bottles of beer.' He said, 'Gosh, I haven't seen a beer in many, many months.' I looked at Dick and he looked at me with a smile, and I handed the Sergeant a bottle. You would have thought he had just won the lottery. A couple of hours after we took off from there we were peacefully cruising along. Dick said, 'There's a flight of Jap fighters at 10 o'clock.' We both sat very still and they didn't see us. To this day I don't know how they missed us. When we got to Middleburg we went through the same exciting landing procedure—up and down—out of control—in control. I still don't know how he managed to get it on the ground. I doubt if very many pilots could have handled it."[1]

Upon arriving at their new location the personnel established their housing. Coconut logs and debris needed removal to make space for the tents. Stewart and Allen found a quick way of accomplishing this necessary chore:

"After landing, we started the hot task of clearing an area for our tent. I saw a bulldozer doing some work not very far away and went over and asked if he could come over and clear a tent area for us. He said 'no' and remarked that he was just too busy. I asked him if he could do it for a couple bottles of beer. He wheeled that thing around and I think he beat me to the tent site."[2]

The detachment commenced operations on August 24, 1944, with routine dusk and dawn patrols and an alert crew on the ground during the hours of darkness. Japanese night intruders flew missions against Middleburg and Mar airfields from bases at Ceram (an island off the southwest coast of New Guinea), as well as Halmahera and Mindanao. The day after the 419th NFS began flying operational sorties Japanese night intruders bombed Middleburg, and according to Captain Warren Slater, the squadron intelligence officer, "The personnel began logging fox hole time." Two Japanese planes dropped anti-personnel bombs on 27 August; however, they missed the 419th NFS bivouac area and flight line. First Lieutenant Robert O'Brien, the 419th NFS ordnance officer, recalled these bombs dropped by the Japanese:

"They were canisters of little anti-personnel bombs the Japanese made by modifying rifle grenades and loading about one hundred or more into a large case. They were designed to pop open in flight and then spread the small grenade bombs over a large area."

Inside the 419th NFS Operations shed at Middleburg Island on January 30, 1945. Note the flying status board listing P-61 Black Widows #324 (pilot Captain Don Dessert), #321 (pilot First Lieutenant Bernard Knott), #302 (pilot First Lieutenant Kenneth Voy), and #305 (pilot First Lieutenant George White) as being operational that day. According to the board Dessert flew to Morotai, where the other 419th NFS detachment was based. The note on the right side of the P-61 list warns "Keep your buzzing at tree top high." *Courtesy of Don Dessert.*

The 419th NFS Operations shed on Middleburg Island. Note the communications van parked alongside the taxi way and *Elusive Susie*, the P-38 Lightning (#302) often flown by Captain Dick Stewart, behind the van. *Courtesy of Don Dessert.*

The Japanese bombs failed to explode during one raid and O'Brien investigated the potential threat they posed lying scattered north of the bivouac area:

"The fuse worked from a small propeller in the tail and had not quite armed during descent. I found that I could stick a straightened paper clip into the hole by the fuse and render it safe. This I did with all I could find and then I dumped them in the nearby ocean."[3]

The next night, a Japanese Aichi D3A Val dive bomber struck the island, causing little damage. Vals were two-seat naval dive bombers that frequently operated from aircraft carriers. In New Guinea, the Japanese utilized these aircraft for night intruder missions from land bases. A Val had a maximum ceiling of 35,700 feet, a range of approximately 1,130 miles, and a speed of approximately 240 to 280 miles per hour. Although slower and capable of less range than the two engine bombers normally employed for intruder work, the Val was more maneuverable if intercepted by an American night fighter. It carried two fixed, forward mounted 7.7 mm machine guns in the wing and a single 7.7 mm machine gun behind the pilot for rear defense by the second crew man. The offensive armament of a Val included the capability of carrying a single 500+ pound centerline bomb and two 66 pound bombs under the wings.

Military personnel have often been able to find humor in the most serious of situations. Flight Officer Joseph Sokolewicz, a Radar Observer (R/O) and squadron special services officer, established a small outdoor theater for the unit, and the men were watching a movie when a Val swooped over the island. Captain Slater casually noted in the monthly historical report:

"The orderly and rapid manner in which the audience attending the movie dispersed when a red alert sounded in the middle of the feature picture on the night of the 28th was indeed commendable."[4]

Another Val appeared the night of 31 August and buzzed the runway from north to south, dropping several small bombs and causing minor damage to the steel matting. Sergeant Jim Girdis, a radar mechanic with the 419th NFS, recalled having to endure 13 red alerts in a single night:

"The Japanese were flying toward and then away from the island all night just to keep us awake. We often had to sleep at the edge of or in our fox holes."[5]

The night fighter crews of the Headquarters Detachment either flew patrols or sat on alert, depending upon the location and level of enemy activity. The pilots

Shower facility. Note how the water flows from the large container along pipes to small collections of shower heads. *Courtesy of Don Dessert.*

419th NFS Laundry Tent. Note the shower area behind the laundry. *Courtesy of Don Dessert.*

sat on strip alert much of the time at Middleburg. The designated aircrew members waited in an alert shack adjacent to their aircraft. They would either sleep, play games, or read. The P-61 mechanics started the engines of the aircraft every hour to keep them warm and ready for an immediate take off. When alerted by Ground Control Interception (GCI) personnel, someone would hit a gong and everyone would rush to their assigned positions. The aircrew members jumped into their waiting aircraft and strapped themselves down. During this process the local anti-aircraft (AA) battery fired three bursts into the air as an air raid warning to the entire airfield. When ready, the P-61 or Lockheed P-38 Lightning raced down the runway and took off to intercept the incoming bogey. Stewart held the squadron record for the shortest length of time in this process. One night only 57 seconds elapsed between the first beating of the gong and when the tires of his Black Widow left the ground.

The Middleburg GCI controller scrambled First Lieutenant William Mackie and his crew, Second Lieutenant Roger Egan (R/O) and Staff Sergeant Frank McCormack (gunner), at 3:03 AM on August 31, 1944, the same night the second Val dive bomber raided the island. The controller vectored the P-61 crew until Egan acquired a fix on an incoming bogey with the on board SCR 720 radar system. Egan informed Mackie that their target was six miles away. Within seconds, the incoming aircraft disappeared from the radar scope. While orbiting the area attempting to re-acquire a radar fix, the P-61 attracted the attention of the local AA battery.

The 419th NFS "MGM Sokolewicz Theater" on Middleburg Island, named in honor of Flight Officer Joseph Sokolewicz, who held the additional duty as the squadron special services officer. The covered projector platform sits to the left of the photo and the screen is on the right. Note the logs doubling as benches which increase in height toward the rear, similar to stadium style seating. One evening a Japanese Aichi D3AVal dive bomber surprised the island during a film at the theater. Captain Warren Slater, the squadron intelligence officer, calmly wrote in his report, "The orderly and rapid manner in which the audience attending the movie dispersed when a red alert sounded in the middle of the feature picture on the night of the 28th was indeed commendable." *Courtesy of Don Dessert.*

419th NFS quarters area on Middleburg Island. The day's laundry hangs on a line between a tent and a tree. Note the fox hole located at the edge of the tent area (with lighter colored sand on its edges and reinforced with logs and sand bags). A second foxhole can be seen behind the barrel at the base of the tree in the foreground. An ambulance sits halfway down the row of tents. *Courtesy of Don Dessert.*

Five American searchlights located the Black Widow above Middleburg and the AA guns began firing at the airplane, oblivious to the fact that it was friendly. Mackie took evasive action and escaped the search light beams without sustaining any damage from the AA barrage.

Mackie continued his patrol without locating any bogeys and turned back to land at the air strip after evaluating the weather conditions. Scattered rain showers continued to pass through the area, and the pilot knew that thunderstorms were moving in from the west. A malfunctioning fuel gauge provided additional justification for ending the search. As Mackie approached the air strip, the GCI controller radioed to warn that the pilot had a bogey on his tail. Mackie immediately cut his aircraft recognition lights, dove to his starboard, and then climbed to 6,000 feet as Egan unsuccessfully sought a radar fix on the unidentified aircraft trailing the Black Widow. The crew discontinued their search and landed at 5:10 AM. After the mission, First Lieutenant John Guzey, the squadron assistant intelligence officer, offered a simple after action statement referencing the GCI controller, "GCI should keep friendly aircraft outside of the AA ring of fire."[6]

Night fighting involved the precise coordination of the GCI controller and the airborne crew attempting to intercept a bogey. The early GCI and airborne radar equipment, when working properly, could baffle inexperienced as well as experienced operators, leading to botched intercepts. While the blip produced on a scope could fairly easily indicate the direction of the bogey from the radar receiver, determining altitude proved a greater challenge. A possible error by the GCI controller may have cost Headquarters Detachment a kill on September 9, 1944. The controller scrambled First Lieutenant Thomas Segrest and his crew, Flight Officer George Walters (R/O) and Private First Class Virgil Bell (gunner), at 3:50 AM and directed them to an altitude of 16,000 feet, where they should begin an SCR 720 radar search for an incoming bogey. Segrest flew to within five miles of the reported location of the unidentified plane and Walters initiated an unsuccessful radar search. The GCI controller then radioed that the Middleburg searchlights had located the bogey at 9,000 feet. Segrest maneuvered his P-61 into a steep dive to intercept the aircraft while it was pinpointed by the searchlights. At the same time, the GCI controller issued a "hold fire" call to the local AA battery. Segrest broke through a layer of clouds to find the searchlights had lost their prey. The crew increased altitude to 12,000 feet and continued a sweep of the area with the SCR 720 radar system. Within minutes, the GCI controller radioed a new intercept vector for a bogey departing the area. However, Walters could not locate the aircraft on his radar scope. The disappointed crew returned to Middleburg and landed at 5:30 AM. First Lieutenant Guzey believed the GCI controller may have provided a wrong initial altitude to the Black Widow crew. If the P-61 crew had been flying at the proper altitude when the searchlights located the bogey, there might have been one less Japanese aircrew returning home that morning.

Radar proved to have challenges in distinguishing between contacts when they were flying closely together. The radar signatures of multiple aircraft could appear to be one blip due to the inability of the early systems to separate the images of planes in close formation, or even at different altitudes but the same general vertical location. Second Lieutenant John Radzieta, an R/O, and his crew were scrambled while at Middleburg to investigate a bogey one night. The GCI controller provided a vector and the P-61 flew to an intercept point. The crew overshot the bogey based on the GCI controller's estimate. Radzieta finally picked up a radar contact with the SCR 720. However, upon investigating the bogey, the P-61 crew discovered an American Catalina patrol aircraft below them. Meanwhile, the actual bogey slipped by them overhead, undetected by the P-61's SCR 720. The bogey bombed Middleburg and scooted back through the AA circle. The P-61 crew could not initiate a chase based on the GCI controller's

Master Sergeant Bill McConnell (center kneeling) and a maintenance team working on a P-61 Black Widow. While the P-61 required regular maintenance to remain operational in the tropical environment, the SCR 720 radar system also needed close attention. According to McConnell, the 419th NFS maintenance Line Chief, the SCR 720 demanded considerable power to operate. "The radar would burn up the generators on board the P-61s and they were not easy to replace." Radar and communications failures contributed to the loss of kills against several bogeys chased by 419th NFS Black Widows. *Courtesy of Dot McConnell.*

Staff Sergeant Leonard Keim, the crew chief of *Old Salty Dog II*, hands a parachute to Captain Don Dessert in the cockpit. Dessert remembered that Keim was so tall that he was the only person who could hand a parachute to the pilot without climbing on something. Radar Observer Flight Officer Lane Thompson is securing his parachute from the top of the jeep in the foreground. Note the two Japanese flags on the aircraft representing two separate crews who downed a Japanese aircraft while flying this particular Black Widow. Also note the radar transmitter on the side of the aircraft below the cockpit (scratched out on the original photo for security reasons).The radar receiving dish sat in a compartment located in the nose of the P-61 Black Widow. *Courtesy of Don Dessert.*

directions due to a severe lack of fuel. The pilot set the P-61 down at the airfield and one engine, starved of fuel, quit before the aircraft could be parked in the squadron area.

Japanese pilots utilized the mountains and valleys around Sansapor to mask their movement during intruder runs on the American base on Middleburg. When radar waves strike a mountain or the ground they bounce back, indicating the presence of the terrain. Any object flying along the terrain would have its radar signature covered by the much greater return of the landmass behind them. New Guinea provided the Japanese with many valleys and mountains to utilize in shielding themselves from American radar. Japanese intruders in the Sansapor area received an additional terrain related force multiplier on their missions. The American crew who set up the local GCI radar station situated it in a very poor location. The radar dish sat less than a mile from a small mountain that cast a 30 mile diameter "dead area" where Japanese aircraft could not be detected due to the terrain blocking the radar waves.

Second Lieutenant James Allen, the squadron chief R/O, remembered the Japanese use of terrain to avoid radar detection:

"We were on patrol when the GCI controller called in that we had an incoming bogey. He vectored us to a location to intercept the bogey. However, the bogey went into the mountain valleys. My radar was working well that night but we couldn't pick him up. The Japanese aircraft emerged from the mountain valleys, bombed Middleburg, and made his get-away."[7]

Allen and Stewart, his pilot, attempted to pursue the Japanese intruder after it struck Middleburg, but then encountered a series of different problems. The GCI controller appeared to be very inexperienced. He kept providing radar fixes to the crew that did not make

419th NFS maintenance tents along the airstrip at Middleburg Island. *Courtesy of Don Dessert.*

A 419th NFS P-61 Black Widow taxiing to the airstrip for a mission.
Note the Operations shack and communications van in the right of the picture. *Courtesy of Don Dessert.*

sense to them. Allen could not pick up any bogeys with his SCR 720, yet the GCI controller kept reporting the bogey was just ahead of the P-61. The American crew made the decision to abandon the chase and return to Middleburg due to a growing fuel shortage issue. The P-61 pilot flew back toward Middleburg from the north and initiated a series of maneuvers to show they were friendly. However, the airfield refused to turn on the landing lights, reporting that they were still under a "red alert." The fuel shortage in the P-61 was now critical, and the crew did not have the time to orbit the area. Stewart commenced his landing sequence, but could not see the end of the runway due to smoke and flames from burning aircraft hit during the Japanese raid. The crew flew across the runway, pulled a 180 degree turn, and quickly came back for a landing out of concern of being shot at by friendly AA situated at the end of the runway.[8]

The Japanese were keenly aware of the capabilities of American airborne and ground-based radar systems and developed several countermeasures to negate these defensive attributes. Many multi-crew Japanese aircraft carried an English-speaking crewmember who listened to known radio frequencies utilized by Allied planes. These Japanese specialists would alert their pilot if the American GCI controller reported their position to a night fighter. At this point, the Japanese intruders would take evasive action to avoid detection by airborne radar of an incoming American night fighter. On many occasions, a GCI controller reported the presence of a bogey that would later disappear and not be found by the night fighter sent in pursuit of it. Many of these Japanese aircraft turned around and returned to their points of origins rather than continue their missions after being spotted by radar. The English-speaking Japanese crewmembers sometimes attempted to redirect the American night fighter by issuing orders contradictory to the GCI controller. While flying out of the Solomon Islands, the 419th NFS reported the following radio transmission between a pilot, the GCI controller, and a Japanese intruder:

Captain Dick Stewart. Note the infantry leggings on the top of his boots. *Courtesy of Dick Stewart.*

GCI controller: "MacDonald from Kiwi. Vector 300 degrees. Over."

419th NFS Pilot: "300 degrees, roger."

Japanese intruder: "Kiwi to MacDonald, orbit Savo."

419th NFS Pilot: "I heard that Japanese S.O.B and am still on 300 degrees."

American night fighter pilots and GCI controllers developed several methods to counter the English-speaking Japanese crewmembers. First, the Americans developed simple word substitution codes to confuse Japanese intruders. For example, the GCI controller might report, "10 beers on 120." The P-61 pilot would then translate this phrase as, "Bogey 10 miles away at a vector of 120 degrees." Naturally, for these word substitution codes to work the CGI controllers and pilots would have to constantly change them. Utilizing word substitution codes allowed First Lieutenant Robert Frenzel to intercept a Japanese Kawasaki Ki-48 Lily on September 11, 1944, and Captain Don Dessert to shoot down a Japanese Mitsubishi Ki-46 Dinah on October 25, 1944. Second, GCI controllers sometimes faked radio conversations.

419th NFS P-61 Black Widow 42-5563 #324 *Old Salty Dog II*. Note the two Japanese flags below the cockpit symbolizing the two kills made by pilots flying this aircraft. *Courtesy of Don Dessert.*

419th NFS P-61 Black Widow 43-8238 #332 *Vivacious Vivian. Courtesy of Jeff Kolln.*

Another view of 419th NFS P-61 Black Widow 43-8238 #332 *Vivacious Vivian. Courtesy of Adam Smith via Jeff Kolln.*

In one method, a GCI controller would talk to the patrolling night fighter pilot on one frequency while a second controller faked a conversation with a non-existent night fighter pilot on a second frequency by changing his voice and answering his own directions. This forced the Japanese night intruder pilot to choose between the two conversations and not always know if he had been spotted on radar. Once, on Guadalcanal, a GCI controller acquired a fix on a bogey 100 miles from the island. Captain Edward Thompson, a GCI controller on Middleburg, finished the tale:

"The controller acted as both fighter and controller on HF radio, called himself in as an airborne night fighter and, changing his voice, acknowledged himself. He vectored himself into position in this manner, eventually saying, 'Bogey 11:00 o'clock, 10 miles.' The bogey immediately turned back up the slot away from Guadalcanal showing a ground speed in excess of 350 m.p.h. No fighter was airborne – yet the Japanese plane took evasive tactics!"

Left to right: Captain Don Dessert (pilot), Second Lieutenant Lane Thompson (Radar Observer), and Staff Sergeant Frank McCormack (gunner). *Courtesy of Don Dessert.*

Japanese intruder pilots quickly learned that violent maneuvers, especially diving, could cause them to disappear from American radar screens. They also employed "window," or aluminum strips thrown from an aircraft to produce false signals on radar. Radar signals struck the mass of aluminum and returned to the receiver dish as though they were bouncing off an aircraft in the sky. A small amount of "window" could give a radar return that seemed to indicate the presence of a non-existent aircraft, while large quantities of the strips could actually mask an entire portion of sky, preventing the radar from seeing behind the curtain of aluminum floating in the air. Japanese intruders frequently employed "window" as part of their tactics when attacking Middleburg. Three or four bogeys would approach Middleburg as a group stretching approximately 100 degrees in the sky. Each plane dropped "window" simultaneously and immediately turned between 90 and 270 degrees through the aluminum while also diving. The turning and diving maneuver made them disappear on radar screens while the aluminum "window" then produced a false return on the scopes. The Japanese aircraft continued to dive to fool the radar and then initiated a steep climb to an altitude of 11,000 to 15,000 feet to make their bomb run. They frequently dropped their ordnance on Middleburg and had flown up to 15 miles before American radar re-acquired them. On 28 August, only three days after the placement of the GCI controller at Sansapor, five Japanese intruders made a coordinated strike on Middleburg utilizing a variation of this tactic. All five aircraft dropped "window," but one continued to drop the aluminum strips to confuse the American defenders and act as a decoy. The other four aircraft then bombed Middleburg and escaped. As they departed the area, attracting the attention of the Americans, the fifth plane sneaked through the radar defenses and also bombed the island.

Not every unidentified aircraft was a Japanese intruder. Allied aircraft carried an Identification Friend or Foe (IFF) device. The IFF transponder emitted a signal to identify the plane as friendly. However, if the device malfunctioned or was not turned on during a flight, the aircraft would be marked as "unidentified" once detected on radar and the GCI controller would scramble a night fighter to intercept it. The GCI controller scrambled First Lieutenant Stewart to intercept an incoming bogey early one morning. Stewart and his P-61 crew took off and headed west of Middleburg. Second Lieutenant Allen, the R/O, acquired a radar fix on the SCR 720 and directed Stewart closer for a visual identification of the aircraft. At approximately 4:00 AM Stewart spotted the bogey and radioed the GCI controller. Stewart was puzzled by what he saw. The bogey had four engines and looked like an American Consolidated B-24 Liberator bomber. However, every time the P-61 leveled near the bogey it would turn away from the Black Widow. Stewart moved in closer and informed the GCI controller the bogey looked like a B-24 but had only one stabilizer and asked, "Should I shoot it down?" Immediately upon asking the question, the P-61 crew saw cigarette lighters and flashlights being turned on and waved around inside the bogey. Stewart reported what he witnessed to the GCI controller who then told him to follow the bogey until sunrise. "If it approaches any Allied base, shoot it down." Stewart followed the bogey, and at sunrise was shocked when the sun revealed an American B-24 with a large hole in the fuselage and a missing stabilizer. The American bomber landed at Biak and Stewart turned around and headed for Noemfoor, where Detachment A

Left to right: Captain Dick Stewart and Flight Officer Jim Allen. Stewart and Allen, his Radar Observer, were vectored onto an unidentified bogey while operating from Middleburg Island. The P-61 Black Widow crew could not positively identify the unknown aircraft and radioed for permission to shoot it down, not realizing they were following a battle damaged B-24 Liberator bomber limping home. The B-24 crew could hear messages on their radio but could not transmit. When Stewart radioed for permission to fire, the crew of the B-24 opened all of their cigarette lighters and turned on their flashlights in a successful attempt to signal the American night fighter tailing them. *Courtesy of AFHRA.*

Japanese Kawasaki Ki-48 Lily bomber. First Lieutenant Robert Frenzel scrambled with his crew, Second Lieutenant George Black (R/O) and Staff Sergeant Cecil Fuller (gunner), during the night of September 11, 1944, in pursuit of a Lily. A turret gun failure prevented an almost sure kill of the Japanese bomber. *Courtesy of Bob Fowble.*

of the 419th NFS was based. He refueled at Noemfoor and returned to Middleburg.

Ordinarily this would have been the end of the story. Stewart returned to the United States with Allen on a troop ship in mid-1945. During the long voyage, many passengers sat around on deck and told "war stories" from their personal experiences. Stewart and Allen were telling others on the vessel about encountering the damaged B-24. Stewart mentioned that several of the men began grinning from ear-to-ear and informed the P-61 pilot that they were the crew he stalked that night. A Japanese plane had rammed the B-24 in mid-air, but the crew managed to keep their bomber flying. The IFF and radio of the bomber were damaged. However, the crew could receive, but not transmit, on their radio. They were listening to the frequency utilized by the Middleburg controllers and heard Stewart ask whether to shoot them down. At that point they opened all of their cigarette lighters and turned on their flashlights in an attempt to signal the American night fighter tailing them. The B-24 crew informed Stewart that they had made a pact to buy the P-61 crew dinner if ever finding out who they were that night as a "thanks" for not shooting them down. True to their word, the B-24 crew took Stewart and Allen "out on the town" when the vessel arrived in San Francisco.[9]

A turret gun failure prevented an almost sure kill of a Japanese Lily bomber two nights after First Lieutenant Segrest's unsuccessful hunt. First Lieutenant Frenzel scrambled with his crew, Second Lieutenant George Black (R/O) and Staff Sergeant Cecil Fuller (gunner), during the night of 11 September. The P-61 lifted off at 1:20 AM and patrolled to the south as directed by the GCI controller, who had lost trace of the incoming bogey. Frenzel was located 30 miles southwest of Middleburg and at 12,000 feet when the GCI controller radioed an intercept vector. The P-61 crew followed the vector until Black obtained a radar fix on a bogey flying six miles away from them. Black directed Frenzel to a position 2,000 feet from the unidentified aircraft and the crew initiated a visual search of the area. The pilot and gunner spotted the bogey and identified it as a Japanese Kawasaki Ki-48 Lily bomber. The Lily generally flew between approximately 300 and 320 miles per hour with a rather short range of 1,419 for a bomber. It was

The 419th NFS lost a P-61 Black Widow during a flight test on September 13, 1944. Both booms on the rear of the aircraft had buckled during a hard landing on 25 August. Following repairs, First Lieutenant Dick Stewart, as the unit test pilot, took her up to verify the plane's readiness to return to combat missions. During a high speed dive, strain on the aircraft warped the Radar Observer's compartment and imploded the Plexiglas cone. The right tail boom and horizontal stabilizer rippled and warped. Stewart managed to bring the Black Widow under control and successfully landed the aircraft, which was then transferred to the 38th Service Squadron for salvage. Although a similar incident happened to the P-61 in this photo, it cannot be verified that this is the same plane Stewart tested. *Courtesy of Bob Fowble.*

normally armed with three 7.7 mm defensive machine guns.

The Japanese intruder pilot made a hard turn and Frenzel followed him. The maneuvering meant Frenzel would have a difficult time operating his fixed 20 mm cannon, so he directed Fuller to track the bogey with the .50 caliber machine gun turret. "The gunner tracked the bogey for approximately 30 seconds but the turret guns refused to fire." Fuller radioed for Black to assume control of the turret guns. A P-61 machine gun turret could be operated by either of three crewmen on the aircraft. Later in the war, many Black Widow crews in squadrons flying in both theaters fought without a gunner. In these cases, the pilot normally fired the machine guns along with the cannon. Gun turret control could be passed to the R/O in the rear compartment as a means of defense against enemy aircraft trailing a P-61. In the case of Frenzel's aircraft, the R/O could not control the gun turret because it became stuck in the vertical position.

Frenzel continued a hard starboard turn as his crew wrestled with the machine gun turret. The P-61 suddenly passed within 30 feet of its prey. Frenzel noted, "We were fired upon as we passed astern and also as we passed over the Lily's nose with undetermined results." The Black Widow flew so close to the Lily that Frenzel reported being able to see the cockpit dials and the Japanese aviator's face. The two planes broke contact and Black lost his radar fix. The GCI controller provided a new vector to intercept the escaping Japanese aircraft; however, the crew could not locate it. Frenzel continued his patrol and landed his plane at 4:10 AM. The P-61 crew would have probably downed the Japanese bomber if the four .50 caliber machine gun turret had been functioning properly.[10]

The Detachment lost a P-61 during a flight test on 13 September. Both booms on the rear of the aircraft had buckled during a hard landing on 25 August. Following repairs, First Lieutenant Stewart, as the Detachment test pilot, took her up to verify the plane's readiness to return to combat missions. During a high speed dive, strain on the aircraft warped the R/O's compartment and imploded the Plexiglas cone. The right tail boom and horizontal stabilizer rippled and warped. Stewart managed to bring the Black Widow under control and successfully landed the aircraft, which was then transferred to the 38th Service Squadron for salvage.[11] The Plexiglas cone at the rear of the R/O's compartment had always been a hazard during

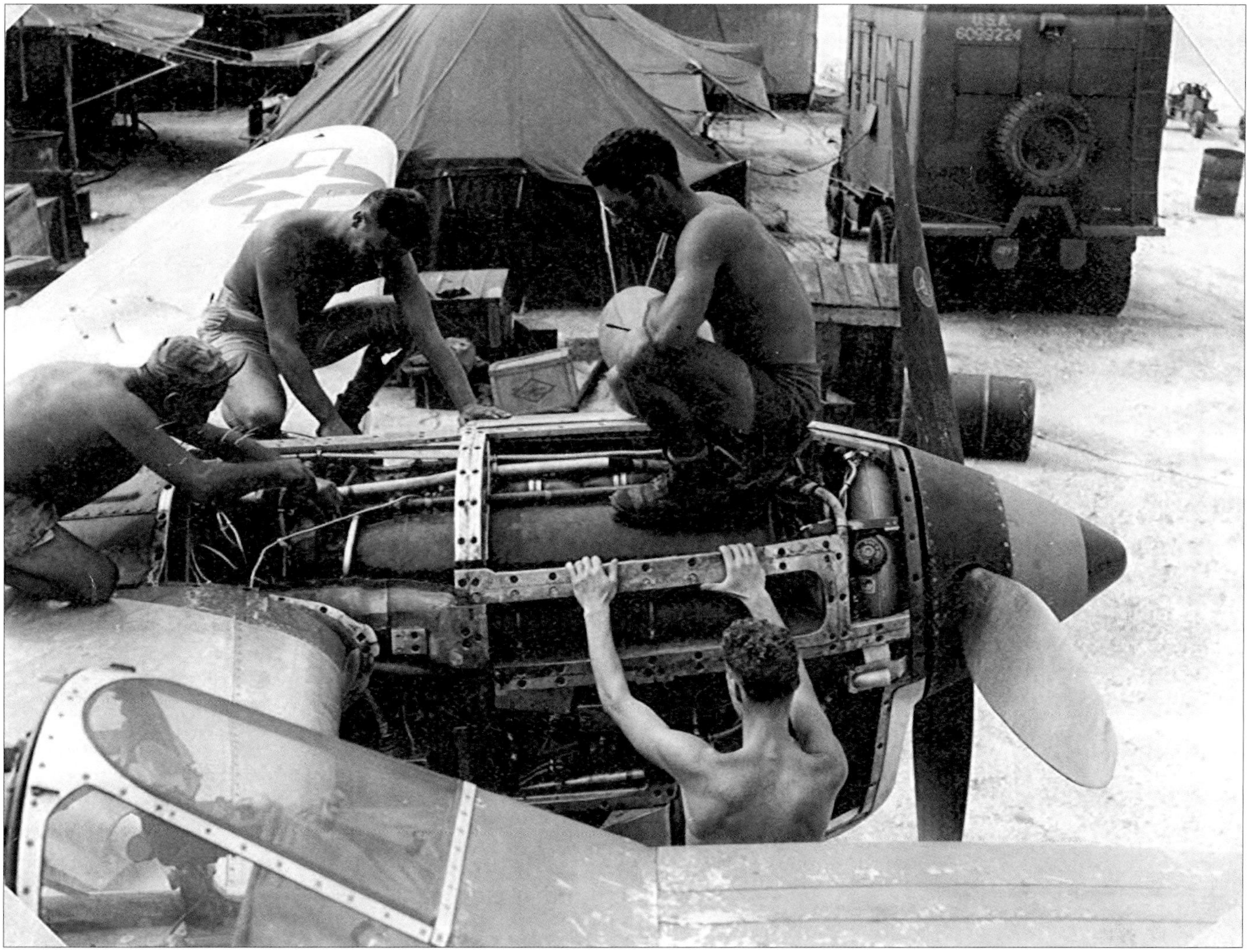

Staff Sergeant Frank Cunningham (kneeling on top of the engine) performs maintenance with his crew on the 419th NFS' P-38 Lighting *Elusive Susie*. The ground crews of the 419th NFS performed near miracles keeping the squadron's aircraft airborne in the tropical conditions of the Southwest Pacific. Sergeant Harold Young noted that fine coral dust proved to be a serious problem in that area of the world. "The coral was really harsh on the aircraft. In particular, coral dust was very abrasive. When we ran the engines on the ground, the propellers would pick up the dust and put little nicks into the blades." *Courtesy of AFHRA.*

high speed dives, and the aircraft generally carried a field modification of a horizontal stabilizer bar running across the inside of the Plexiglas. As the pressure and strain of a dive began to collapse the Plexiglas, the bar would prevent the two sides from collapsing into the middle of the R/O's compartment. Although the problem was noted during initial testing of the aircraft, the first operational blowout occurred in the European Theater of Operations. First Lieutenant Herman Ernst of the 422nd NFS dove his P-61 to gain speed in order to intercept a German V-1 Buzz bomb over Great Britain on July 15, 1944. The Plexiglas cone on the R/O's compartment imploded due to the pressure of the dive and forced Ernst to break off his pursuit of the V-1.

Flight Officer John Radzieta related a warning Middleburg received from Tokyo Rose:

"Middleburg's elevation was only about twelve feet and that was at the runway. Tokyo Rose kept warning us in her broadcasts, 'You boys on Middleburg are in for a big surprise.' She even gave us a specific date which I do not remember now. We grew concerned. We were no fools and made preparations. We dug deep foxholes and covered them with coconut logs to protect ourselves against anti-personnel bombs and shrapnel. We made sure we had gas masks that worked and emergency rations and water in our foxholes. We converted helmets from toiletry basins back to their original purpose of head protection. We were issued additional ammo for our small arms in case we were invaded by the Japanese. We doubled our planes and crews on alert. Many individuals found excuses as to why they had to be off the island on that day. We sweated out the night, knowing that our

Maintenance on a P-61 Black Widow engine. Northrop powered the Black Widow with two Pratt and Whitney R-2800-10 supercharged engines with 2,000 hp. Later P-61s carried the R-2800-65 engine, which provided an additional 250 hp. Note the caricature of Old Salty Dog and the single Japanese flag below the cockpit. *Courtesy of Don Dessert.*

foxholes were just a leap away from our tents. At sunup I let out a big roar. Water was in our tent and it was high, within an inch or two of having my air mattress afloat. Shoes, coconuts, and dead crabs were floating in the tent, and the entire area was covered by water. The foxholes constructed by toil and sweat were full of water. It was a high tide and Tokyo Rose knew that it was coming."

Flight Officer Jim Allen and his pilot, First Lieutenant Dick Stewart, made a similar discovery that morning. Allen recounted:

"The bulldozer operator who dug out our tent site for some beer dug out too much dirt. When we had the high tide, Dick and I woke up with water up to the bottom of our cots and our shoes and other things floating around, and a bunch of guys standing there laughing at us."[12]

Food quality remained an important morale factor for the night fighters. Mail from home rivaled food when it came to morale. Flight Officer John Radzieta emphasized this point for the squadron personnel on Middleburg:

"There was one thing that you could not mess with. It was 'mail call.' I can remember when we got to Middleburg, a ship with supplies was anchored off shore. The ship had to be offloaded and moved out before dark. They did not want that ship to become a target and sunk in the harbor. Every spare person became a stevedore or baggage handler. This worked fine until the mail plane came. Some wanted to take a break for mail call. It sounded like a small mutiny until the First Sergeant mustered the troops and redefined the task, pointing out that no mail would be distributed until the cargo was

Left to right: Captain Don Dessert (pilot), Second Lieutenant Lane Thompson (Radar Observer), and Staff Sergeant Frank McCormack (gunner). *Courtesy of Don Dessert.*

removed from the vessel. It was off loaded in record time."[13]

Japanese intruders struck the bomber field at Mar during the morning hours of October 1, 1944.[14] Although two P-61s and a P-38 were airborne, they could not locate the Japanese aircraft. Sergeant Jesse Gardner, a 419th NFS operations clerk, vividly remembered watching the air raid from across the channel on Middleburg:

"We stood out in the open and watched the Japanese aircraft come in low over the field, occasionally in a searchlight, ack-ack tracers shooting at them, and the flash of their bombs as they struck. Ten minutes later, a delayed-action bomb which had landed in an ammunition dump on shore went off and caused a second red alert."[15]

Hours after the Japanese attack, Headquarters Detachment suffered two major incidents resulting in one death and the complete loss of two aircraft. Staff Sergeant Fred Straub, a flight chief, died when a P-61 cannon malfunctioned due to a circuit shorting out. Technical Sergeant Eddy Ballard, the crew chief of the aircraft, reported that the Black Widow experienced generator problems, but these were corrected with assistance from the 38th Service Squadron. The day of the accident, the maintenance personnel were working on the brakes in preparation for a night mission. Ballard left the area in the early afternoon and his maintenance assistants remained with an engineering team to complete the aircraft preparations prior to a combat mission. The team, consisting of Straub, Technical Sergeant Leo Hays, and Corporal Richard Bauer, was making a routine voltage check on the electrical circuit of the aircraft when they shorted out the cannon circuit breakers. In this unusual incident the guns fired despite the

First Lieutenant Michael Burlingham, 419th NFS Engineering Officer. Burlingham and Corporal Billie Cook, a mechanic, received shrapnel wounds from the exploding shells of a malfunctioning P-61 Black Widow 20 mm cannon. A maintenance team, consisting of Staff Sergeant Fred Straub, Technical Sergeant Leo Hays, and Corporal Richard Bauer, was making a routine voltage check on the electrical circuits of the aircraft when they shorted out the cannon circuit breakers. In this unusual incident, the guns fired despite the presence of four safety features on the aircraft to prevent accidental operation of the weapons. Four high explosive shells exited the cannon and struck the activating levers, braces, and doors of the front landing gear. Straub died in the accident that wounded Burlingham and Cook. Hays and Bauer received burns. *Courtesy of Leslee Lukas Mays.*

The air and ground crew members of 419th NFS P-61 Black Widow 42-5497 #322 *Dawn*. First Lieutenant Ralph LeVitt named the aircraft for his wife. #322 is the aircraft in which an electrical malfunction caused the cannon to fire, killing Staff Sergeant Fred Straub. Note the small square below the pilot's cockpit. This is a Japanese flag for LeVitt's kill of a Japanese Mitsubishi G4M Betty bomber on January 9, 1945. *Courtesy of Don Dessert.*

presence of four safety features on the aircraft to prevent accidental operation of the weapons. Four high explosive shells exited the cannon and struck the activating levers, braces, and doors of the front landing gear. The rounds exploded, sending shrapnel into the nose wheel well and starting a fire that quickly reached the cockpit where Straub sat. Straub died in the fire and Hays, the line chief, and Bauer, the assistant crew chief, suffered first and second degree burns. First Lieutenant Michael Burlingham, the engineering officer, and Corporal Billie Cook, a mechanic, received shrapnel wounds from the exploding shells. All of the injured personnel except Stroud recovered. An investigation, conducted by First Lieutenant Dessert, recommended that the first four rounds of each P-61 cannon should be loaded with ball ammunition rather than high explosive shells. In the unlikely event of another accident, ball ammunition, which is essentially a giant bullet, would not explode. The death of Staff Sergeant Straub and the injuries to the other four individuals resulted from the exploding shells as they struck the front nose gear compartment, an area retracted during flight. An accident involving ball ammunition might have resulted in no injuries to the ground crew around the aircraft.[16]

In the unit's second accident of the day, First Lieutenant John McGhee ditched one of the detachment's two P-38s at sea. Both P-38s were up that day on a routine intruder patrol flown by McGhee and Second Lieutenant Guy Scheetz in the second aircraft. The pair searched for any Japanese activity along the southern coast of Waigeo Island. McGhee dropped down to 500 feet and passed twice over a village. On his second pass the left engine ceased operating. McGhee immediately turned his aircraft out to sea, trimmed the plane, and changed fuel tanks. The engine restarted momentarily and then cut out again as the right engine began giving the pilot problems. McGhee cut his engines and ditched the P-38 in the ocean. The plane sank immediately, but the pilot escaped with only minor cuts and bruises. Scheetz provided cover until a Consolidated PBY Catalina Dumbo arrived to retrieve McGhee and fly him back to Middleburg. The Catalina was a two-engine

Four P-61 Black Widows of the 419th NFS lined up at Middleburg Island with 42-5508 #313 in the foreground. Note the blackened areas on the side of each aircraft. Originally, early photos of the P-61s were supposed to have the radar emitters blackened out for security purposes. Some photographers followed the rule while others ignored it. Obviously the squadron member who made this photograph took the time to blacken each plane's radar emitter. *Courtesy of Don Dessert.*

patrol bomber/amphibian frequently utilized for rescues and long range reconnaissance. Although the P-38 was obviously a total loss, McGhee was in good shape despite the mishap. The incident left Headquarters Detachment with only one P-38. Tomorrow would be a better day.

First Lieutenant Don Dessert bagged a Japanese Mitsubishi Ki-46 Dinah night fighter on October 2, 1944. Dinahs were normally utilized in reconnaissance and night interceptor roles and reached speeds just over 390 miles per hour with a maximum range of approximately 2,485 miles. These aircraft could carry a variety of offensive and defensive weaponry, including machine guns and cannon. Dessert, with Second Lieutenant Lane Thompson (R/O) and Staff Sergeant Frank McCormack (gunner), flew a routine patrol 40 miles west of Middleburg. At 12:40 AM the GCI controller radioed to warn of two bogeys approaching the island from the south. The intruders were emerging from the mountains of New Guinea and had apparently been utilizing them to mask their aircraft from American radar. Dessert banked his Black Widow and initiated a hunt for the bogeys over New Guinea. Thompson acquired a radar fix on one bogey approximately 30 miles south of Middleburg and provided his pilot with an intercept vector.

Another view of 419th NFS P-61 Black Widows at Middleburg Island. Note the New Guinea mainland across the water. *Courtesy of Don Dessert.*

Captain Don Dessert (pilot) sits inside the jeep. Sitting on the hood are left to right Second Lieutenant Lane Thompson (radar observer) and Staff Sergeant Frank McCormack (gunner). This P-61 Black Widow team downed a Japanese Mitsubishi Ki-46 Dinah near Middleburg Island in October 1944. Dessert wrote in his mission report, "I ordered my gunner to cut loose with the turret .50s and we both opened fire. I fired a 3-4 second burst with the four 20mm cannon and my gunner fired a second longer tracking the target while I was forced to turn to starboard to avoid disintegrated pieces of the enemy aircraft flying rearward. The initial burst raked the Dinah with a fusillade of direct hits along the wing and fuselage causing the Dinah's port engine to immediately burst into flame. The Dinah went into a spin to port and we saw him crash, splatter, and burn approximately 20 miles southeast and inland from Cape Sansapor at 0054 hours." *Courtesy of Don Dessert.*

Dessert reduced power to the P-61's engines as he approached within 6,000 feet of the unidentified aircraft. At 3,000 feet Dessert and Thompson spotted their prey, and at 400 feet positively identified the bogey as a Japanese Dinah. Visibility that night was excellent, with a full moon to guide the Black Widow crew as they crept closer to the Dinah. Dessert reported:

"I ordered my gunner to cut loose with the turret .50s and we both opened fire. I fired a 3-4 second burst with the four 20mm cannon and my gunner fired a second longer, tracking the target, while I was forced to turn to starboard to avoid disintegrated pieces of the enemy aircraft flying rearward. The initial burst raked the Dinah with a fusillade of direct hits along the wing and fuselage, causing the Dinah's port engine to immediately burst into flame. The Dinah went into a spin to port and we saw him crash, splatter, and burn approximately 20 miles southeast and inland from Cape Sansapor at 0054 hours."

A Navy destroyer crew with night vision glasses confirmed the kill. The crew fired 100 rounds of .50 caliber and 80 rounds of 20 mm ammunition to down the Dinah.

Thompson made radar contact with the second bogey and directed Dessert to alter the direction of the aircraft for a second intercept. Dessert

Staff Sergeant Frank McCormack turns the upper machine gun turret of a P-61 Black Widow from his gunner's position located immediately behind and above the pilot's compartment. McCormack, with his pilot First Lieutenant Don Dessert and Radar Observer Flight Officer Lane Thompson, downed a Japanese Mitsubishi Ki-46 Dinah in October 1944.This photo illustrates how Northrop placed the upper machine gun turret behind both the gunner and pilot in order to eliminate temporary loss of night vision due to the flashes of firing guns in an engagement.The upper turret housed four 50 caliber machine guns that could be electrically fired by either of the three crewman on a P-61. *Courtesy of Don Dessert.*

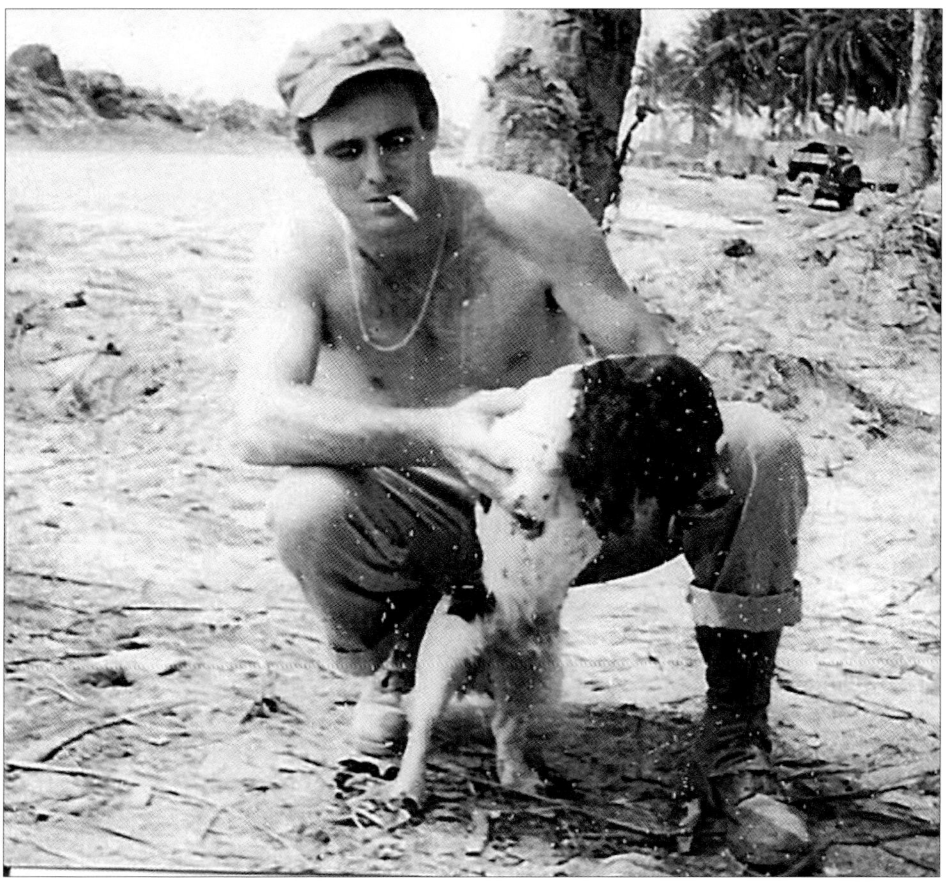

Captain Don Dessert with Old Salty Dog. One of the aircrew members brought the dog back from leave in either Australia or New Zealand. Dessert named his P-61 Black Widow *Old Salty Dog II* in honor of him. *Courtesy of Don Dessert.*

unsuccessfully chased the second bogey through the mountainous region of New Guinea. After landing, the pilot surmised that the pair of bogeys may have been planning a strategy to distract the American night fighter they knew would be on patrol. While one plane decoyed the P-61 into the mountains where radar would become nearly useless, the second bogey would then bomb Middleburg. The second plane they chased may have been the decoy plane and did not realize that the lead bomber had been downed just minutes before being intercepted by the Black Widow. Although the Headquarters Detachment cheered the exploit of Dessert and his crew, there was a degree of sadness among some of the personnel. Old Salty Dog, a small "Heinz 57" pup brought back from Australia, died the same day. In tribute to the mascot, Dessert named his P-61 "Old Salty Dog II."[17]

The squadron lost Captain Howard Daniel, one of the former Royal Canadian Air Force veterans, on 18 October. Daniel and his crew, Second Lieutenant Harold Ozmun (R/O) and Sergeant William Yahn (gunner), were on an intruder mission bound for Waigeo Island when they disappeared. They took off from Middleburg at 4:45 AM and no one received a radio report from them after that time. The base controller, GCI operator, and homing station attempted to locate the missing P-61 but were unsuccessful. At 7:00 two P-61s were dispatched to search the area covered by the mission. Additional searches by P-61s and P-38s failed to locate the plane, any downed crew members, or debris on the ocean over the next two days. The three were declared missing in action. The loss of the crew was a savage blow to the morale of the unit. Captain Daniel owned a small dog that he had acquired in either Australia or New Zealand. As a tribute to their friend, the other members of the squadron arranged for the dog to be transported to the United States and delivered to Daniel's parents, who cared for the canine.

The Headquarters Detachment's second P-38 was damaged during a taxiing incident on October 26, 1944. The right brake failed with First Lieutenant Stewart in the cockpit:

"I was taxiing down the edge of the strip, getting ready for takeoff when the ship started veering to the left. I tried to turn the ship with the throttle but the right brake didn't hold and the ship wouldn't turn in spite of full throttle. The nose wheel hit a large coral rock and buckled."

Sergeant William Yahn. In October 1944 Yahn and his fellow crew members Captain Howard Daniel (pilot and one of the former Royal Canadian Air Force veterans of the 419th NFS) and Second Lieutenant Harold Ozmun (Radar Observer) disappeared while on a P-61 Black Widow intruder mission bound for Waiego Island. Searches by P-61s and P-38s failed to locate any debris from the missing plane. All three were officially declared missing in action. *Courtesy of John Radzieta.*

The accident bent both propellers and slightly damaged the nose area of the aircraft.[18]

First Lieutenant Guy Scheetz encountered an interesting problem while attempting to intercept a Japanese intruder on November 9, 1944.[19] Scheetz and his crew, Second Lieutenant William Gillette (R/O) and Staff Sergeant Leonard Sherman (gunner), began their patrol at 3:15 AM. Within minutes of departing Middleburg, Gillette reported the SCR 720 radar system as inoperable. The R/O conducted a routine check of the fuses and voltage to correct the problem. However, nothing seemed to work that morning. Scheetz continued the patrol as Gillette worked to solve the radar problem. The pilot reached a point approximately 50 miles west of Middleburg and reported the radar problem to the GCI controller, who in turn informed the P-61 crew that an unidentified aircraft approached their position. Scheetz recalled:

"I told the fighter controller to send up the scramble ship immediately. Fighter control replied there was not enough time to get off a scramble ship...I requested that GCI vector me onto the bogey in the hope that I would run onto him and get a visual."

The GCI controller provided the P-61 crew with a series of three vectors and directed them to decrease altitude to 8,000 feet. The controller relayed the bogey's new position at four miles and informed Scheetz to reduce speed. At this point, the Black Widow crew was on their own because the GCI controller could not direct them any closer based on the image scale of his radar system. Normally, the GCI controller would bring a P-61 within four to five miles and the R/O would then utilize the SCR 720 to bring the pilot within visual range of the unidentified aircraft. Scheetz continued to change altitudes to provide his gunner and R/O opportunities to visually scan the night sky.

The crew finally spotted the bogey, flying level and displaying its navigation lights. The Japanese aircraft was single engine with elliptical wings. Scheetz attempted to line up the target because this particular P-61's gun turret could not rotate. Thus, both the machine guns and cannon were fixed, requiring the pilot to adjust the flight path of the Black Widow in order to aim the weapons. As the pilot adjusted his flight path, the P-61 could not be slowed enough to remain behind the bogey. Scheetz reduced speed to 110 miles per hour, just barely above stall speed, and the Black Widow still flew past the Japanese aircraft:

"After zig-zagging, slipping, and stalling my airplane trying to stay behind the target and below it at the same time, the target disappeared into a cloud bank. It was quite evident that we had not been seen, as the target continued to show navigation lights and never changed course or attempted any sort of evasive action."

The crew returned to Middleburg and landed at 5:15 AM. Scheetz reviewed Japanese aircraft recognition charts in the squadron intelligence office and identified that morning's bogey as a Val dive bomber. Scheetz confirmed the aircraft as a Val by not only its shape, but also its exhaust flame pattern. Night fighter squadrons maintained a special set of charts that illustrated aircraft not only by shape but also by their exhaust flames, which were visible at night.[20]

First Lieutenant Bernard Knott also experienced a failure of his SCR 720 radar unit the next morning. Knott and his crew, Flight Officer Peter Ridosh (R/O) and Staff Sergeant Robert Powell (gunner), were scrambled at 3:30 AM to intercept three incoming bogeys. After climbing to 12,000 feet and vectoring to 150 degrees, Ridosh informed Knott their SCR 720 was malfunctioning. The situation worsened when the crew learned not only were there three bogeys in the area, but also seven friendly C-47 transport planes. Since the SCR 720 had failed, the GCI controller directed the P-61 to orbit and wait to see if the Middleburg searchlights could locate any of the bogeys. The controller would then provide a vector for Knott to move in for a kill.

At 4:30 AM the GCI controller radioed the presence of multiple bogeys and directed the P-61 onto an intercept vector. Knott and his crew scanned the night sky for their prey. Knott remembered:

419th NFS P-61 Black Widow 42-39473 #330 *Night Vision*. Before beginning a training mission in Florida, the night fighter crews donned goggles with dark lenses to help their eyes adjust to night conditions. While many waited until a half hour prior to their mission to put on the goggles, others kept them on for longer periods of time. According to First Lieutenant Dick Stewart, crews sometimes played cards while waiting to begin a training mission. Trying to see the hearts and diamonds when playing cards in red-lens goggles proved quite challenging. *Courtesy of Jeff Kolln.*

"My gunner and I saw a green navigation light about 1,000 feet below us and approximately 2,000-3,000 feet off to the port side. This, GCI said, was one of the bogeys, so I went into a diving turn to the port side. Fighter Control cut in at this point and said these planes were not the bogies, but they were C-47s going through the area." Knott pulled out of his dive and leveled his aircraft. "During the next few minutes, Fighter Control and GCI were both contradicting each other as to the identity of these planes. This only served to confuse me and my crew all the more."

Knott continued to trail the bogey displaying the navigation lights:

"While astern of the leading bogey showing navigation lights my gunner spotted two other planes, one to port and one to starboard in a scattered V formation, both trailing planes at a lower altitude than the leading plane."

The P-61 broke off contact in accordance with the directions of Fighter control and landed:

"After landing and talking the situation over with my gunner, who did get a slight glimpse of the lead plane, he described to me what is the exact characteristic of the wing of a Japanese [Mitsubishi Ki-21] Sally bomber. Due to the confusion caused by GCI being contradicted by Fighter Control as to the identity of the planes we didn't have much chance to get a positive identification because Fighter Control had told me to pull away and leave the planes alone."[21]

The Sally is the type of bomber downed by Captain Al Lukas of Detachment A near Noemfoor in August.

A very frustrated Knott remarked after the incident he should have been allowed to acquire a positive identification of the aircraft before being ordered out of the area by Fighter Control. Sergeant Jesse Gardner, the air operations clerk, worked on the line that morning and expressed the exasperation of the entire unit when Knott returned, "We were practically crying on the line, for Knott would have been sure of getting one of them, or possibly two."[22] Knott, 419th NFS commander Major

Newly promoted Captain Dick Stewart scored the only confirmed 419th NFS kill in a P-38 Lightning when he downed a Japanese Mitsubishi G4M Betty bomber over Middleburg Island in December 1944. Stewart spotted the bogey, which was making a slight starboard turn toward his P-38. "I went in on a head-on pass and fired a two-second burst when the bogey filled my sights. I observed tracers striking the right engine and right inboard wing section between the fuselage and engine causing an explosion and burning. I then pulled up to avoid a collision with the burning plane which narrowly missed me." *Courtesy of Dick Stewart.*

Mitsubishi G4M Betty bomber. The Betty, carrying up to seven crewmembers, could attain ranges between 2,262 and 3,192 miles depending upon the configuration of the aircraft. It often carried three 7.7mm machine guns and a single 20 mm tail mounted cannon for defensive purposes. Newly promoted Captain Dick Stewart scored the 419th NFS' only confirmed kill in a P-38 Lighting in December 1944 when he brought down a Betty bomber over Middleburg Island. Sergeant Jesse Gardner of the 419th NFS witnessed Stewart's Middleburg aerial duel while working the night shift and wrote in his diary, "We saw the whole thing…we saw the plane explode when it was hit and fall straight down, burning all the while, into the sea. Quite a sight. We're still picking up shrapnel over the island from the AA fire." *Courtesy of Bob Fowble.*

Shulmistras, First Lieutenant Mackie (Headquarters Detachment Operations Officer while First Lieutenant Dessert was on leave in Australia), and the GCI controller were directed to report to Brigadier General Earl Barnes, the Commander of 13th Fighter Command, the following afternoon to discuss the incident and develop a better set of procedures for night fighter coordination between the GCI controller and Fighter Control.

The Headquarters Detachment lost a P-61 during a landing on December 19, 1944.[23] Second Lieutenant William Rees, a newly arrived replacement pilot, with Second Lieutenant George Christenson (R/O) in the back, landed at 9:40 AM. Rees noted:

"I came on the approach at 150 mph, lowered the flaps and slowed down to 110 mph. About halfway down the approach I cleared the engines, leaving on a little power to pull me up into the field. When I saw that I was going to make the runway I cut the throttles and started to flare out about 75 feet above the runway. The glide wouldn't break. The plane hit on the nose wheel and collapsed it. I cut the engine switches and when they stopped, I opened the top escape hatch and climbed out of the aircraft."

Both men were not injured in the accident, but the P-61 was sent to the 38th Service Squadron for salvage.[24]

Newly promoted Captain Stewart scrambled in a P-38 at 9:15 PM on December 28, 1944, in response to an incoming bogey that was nearing the AA circle. Stewart climbed to 11,000 feet and flew out to a position seven miles east of Middleburg and orbited, waiting to see if the searchlights would locate a bogey. Four searchlights formed a cone around which AA shells began bursting. Stewart banked his P-38 and approached the searchlights at 400 miles

In the Jack Fellows print "*Elusive Susie*," Captain Dick Stewart bags a Japanese Mitsubishi G4M Betty bomber over Middleburg Island. Stewart flew directly toward the oncoming Betty and shot down the bomber as Allied forces on the ground watched with excitement. *Courtesy of Jack Fellows.*

per hour. As the plane crossed the coastline the searchlights disappeared. Stewart withdrew to his original position and remained there until 11:15 PM, when he returned to Middleburg and landed without spotting any bogeys.

Two nights later Stewart applied the same tactics and brought down a Japanese Mitsubishi G4M Betty bomber. The Betty, carrying up to seven crewmembers, could attain ranges between 2,262 and 3,192 miles depending upon the configuration of the aircraft. It often carried three 7.7mm machine guns and a single 20 mm tail mounted cannon for defensive purposes. The plane could deliver up to approximately 2,200 pounds of bombs. The GCI controller alerted the unit of an incoming intruder and Stewart was airborne in the P-38 at 12:15 AM. He flew to a position five miles east of Middleburg and orbited at an altitude of 12,000 feet while GCI tracked the bogey as it approached from the south at a distance of 25 miles. At a range of five miles the controller directed Stewart onto a 120 degree heading for an intercept. Since the P-38 did not carry onboard radar Stewart had to rely on a visual sighting of the incoming aircraft, with or without the aid of ground based searchlights. As he approached the approximate position of the bogey, Stewart remembered seeing the sweeping searchlights and AA tracers rising skyward from Middleburg. The pilot initiated a shallow dive to enter this area of activity. Stewart spotted the bogey, which was making a slight starboard turn toward the P-38:

"I went in on a head-on pass and fired a two-second burst when the bogey filled my sights. I observed tracers striking the right engine and right inboard wing section between the fuselage and engine, causing an explosion and burning. I then pulled up to avoid a collision with the burning plane which narrowly missed me."

Stewart bagged a Japanese Betty bomber that early morning while flying through the American AA barrage from below. He remembers calling the GCI to request a halt in the AA firing. However, squadron members on the ground clearly recall watching the aerial show above them and note that the AA guns were still firing as Stewart engaged the Betty. The AA gunners also claimed the aircraft, but further review of the incident resulted in the kill being awarded to Stewart. Although recorded as Stewart's first and only official aerial victory, the pilot had downed another Japanese Betty bomber, although unclaimed, while flying P-38s in the Solomon Islands with the 419th NFS.[25]

Sergeant Gardner witnessed Stewart's Middleburg aerial duel while working the night shift and wrote in his diary:

"We saw the whole thing...we saw the plane explode when it was hit and fall straight down, burning all the while, into the sea. Quite a sight. We're still picking up shrapnel over the island from the AA fire. AA tried to claim the Jap but Stewart will probably get credit for it."[26]

Stewart's feat is the only 419th kill made by a P-38. Although Captain Van Bibber and at least two other squadron pilots had downed Japanese aircraft while flying P-38s, their victories occurred when they flew with Detachment B, 6th NFS, in 1943 before the arrival of the 419th NFS.

Throughout this period the Japanese continued to move men and equipment onto New Guinea from other bases by camouflaged barges at night. Allied command authorities requested the 419th NFS Headquarters Detachment to patrol along suspected landing points prior to dawn in an attempt to catch the barges while they were vulnerable unloading or offloading adjacent to the shore. Captain Stewart with his regular R/O, Second Lieutenant Allen, conducted one of these patrols when the pilot thought he saw a

Captain Dick Stewart (in the cockpit) talks with Captain Warren Slater, the 419th NFS Intelligence Officer, prior to a mission. Stewart, with his regular Radar Observer, Second Lieutenant Jim Allen, thought they spotted a Japanese barge early one morning near Middleburg Island. In a rare move, Stewart called for Allen to assume control of the upper turret guns from his rear compartment. Allen forgot that the guns had to turn with the turret to the rear of the aircraft before they would line up with his gun sight. However, they would fire if triggered. Allen commented, "I sprayed bullets everywhere as the guns turned back towards the rear of the aircraft. I didn't get too close to the target on that run!" The rising sun illuminated the area better and the crew realized they were trying to engage a large log floating offshore. *Courtesy of AFHRA.*

Second Lieutenant William Gillette standing in front of the living quarters area. Note the laundry on the line running from the tent near the tree. Tent flaps remained up to catch any breeze blowing across the island. *Courtesy of Don Dessert.*

barge in the water below them. As the P-61 dove on the target the plane encountered a pocket of turbulent air, shaking the aircraft and making it impossible for Stewart to aim the fixed 20 mm cannons. He called for Allen to fire the turret-mounted machine guns as they pulled up and over the target. In order for the pilot or the R/O to operate the machine guns the gunner had to release his triggering device. As long as he held it the other two crewmen could not operate the machine guns. The gunner released the trigger as Allen turned his chair in the R/O compartment, aimed the guns, and began firing. However, he forgot that the guns had to turn with the turret to the rear of the aircraft before they would line up with his gun sight. However, they would fire if triggered. Allen commented:

"I sprayed bullets everywhere as the guns turned back towards the rear of the aircraft. I didn't get too close to the target on that run!"

Stewart and his crew pulled up and made another pass on the object below them. The sun was rising, and the crew could now see that all of their efforts had been wasted on a large log floating off shore.[27]

The turret mounted machine guns of a P-61 were designed with a common safety feature to prevent the accidental shooting of a gunner's own aircraft as he tracked a bogey. At every angle where bullets would strike the propellers or rudders of the Black Widow a simple fire interrupter switch would prevent the machine guns from firing. The forerunner of this system was developed during World War I to ensure machine gun bullets would not hit the propeller mounted just past the end of the barrels. The machine guns on a P-61 could not be fired lower than 30 degrees below the horizon in order to protect the fuselage and wings of the aircraft from its own friendly fire. However, P-61 crews did discover that Northrop forgot to take into account one issue when

419th NFS P-61 Black Widow 42-5563 #324 *Old Salty Dog II. Courtesy of Don Dessert.*

designing the Black Widow. Gravitation forces tend to flex aircraft wings during steep dives. Several P-61 pilots discovered this error when they returned to their bases to find bullet holes from their own machine guns in their wing tips.

December witnessed the initial preparations for the detachment to push beyond Middleburg. On 21 December two 419th NFS P-61s were transferred to assist Detachment A on Morotai Island, located approximately halfway between New Guinea and the Philippine Islands. The pace quickened the following month. The 13th Air Force's second complete night fighter squadron, the 550th NFS, arrived from California to replace the 419th NFS on Middleburg. The 550th NFS became operational the previous month and established its main headquarters at Hollandia, on New Guinea's mainland. A detachment of the 550th officially relieved the 419th NFS of its night fighter defense of Middleburg on 15 January. Five days later Major Shulmistras, the 419th NFS commander, transferred to Morotai, where he assumed command of Detachment A as well as the entire 419th NFS, while Captain Don Dessert became acting commander of Headquarters Detachment on Middleburg. The last two 419th NFS P-61s departed Middleburg on February 3, 1945, and flew to Morotai. The ground personnel of the 419th NFS moved from the island to Sansapor on 9 February and boarded LST 619 with the 1034th Signal Company on 12 February. The vessel formed a convoy with seven other LSTs, a destroyer, and a submarine chaser. This group of vessels merged with another to form an even larger convoy of over 100 naval craft days later. Many of the 419th NFS personnel were forced to sleep on the deck of the LST due to a shortage of space within the vessel. Despite naval surface and air protection, a Japanese submarine sank the *USS Renshaw*, a destroyer providing flank cover for the convoy. The remaining vessels arrived safely on Mindoro Island, in the Philippines, on February 22, 1945. Every member of the 419th NFS itched to move ashore and get back into action as a unit.

The 550th NFS designated its men and aircraft on Middleburg as Detachment C. The 550th NFS' Detachment A remained at Hollandia, on New Guinea, and Detachment B consisted of only ground personnel located on Morotai. With the war passing beyond New Guinea and to the Philippines, Middleburg found itself in calmer waters and it was time to lower the night defenses the 13th Air Force mounted over the island. The 550th NFS maintained one P-61 and crew on the flight line in an alert status each night and a second Black Widow and crew on standby alert. The Detachment made only two night

419th NFS P-61 Black Widow 42-5516 #317. The 419th NFS later transferred this aircraft to the 421st NFS where she was named *Dame de la Nuit.* *Courtesy of Don Dessert.*

scrambles after the departure of the 419th NFS. On 9 February Second Lieutenant David Borden piloted a mission in response to a report of a submarine in the area. The next night First Lieutenant Newell Witte scrambled in his P-61 following a second report of a submarine. Neither pilot located a submarine.[28]

On 14 February the 13th Air Force ordered Detachments A and C of the 550th NFS to join Detachment B on Morotai to relieve the 419th NFS as it moved forward to the Philippines. Two Black Widows from the 550th NFS remained behind, and their air crews were temporarily attached to the 42nd Bomber Command at Sansapor. The two P-61s carried out daily dawn and dusk patrols. Following an attack on the radio station at Cape Sansapor, the 550th NFS Black Widows were dispatched to strafe Japanese troops retreating from the area. The 167th Infantry Regiment provided an observer for each aircraft on the mission. One of the aircraft, piloted by Second Lieutenant David Borden with First Lieutenant Irby as his R/O, departed at 6:20 AM and was over the target area within 15 minutes. Each plane made two strafing runs in the areas designated by the observers. Although the pilots could not see any results from their 20 mm cannon, the observers noted that all of the shells landed in the designated target area. Both aircraft returned by 7:00 AM.[29]

The 419th NFS and 550th NFS participated in the offensive to rout the Japanese from the Philippines. However, this narrative will first trace the 13th Air Force night fighter missions in defense of Morotai before examining the 419th NFS and 550th NFS operations in the Philippines. The Morotai chapter will pick up the arrival of the 550th NFS in Hollandia and its operational activation in January 1945.

6

Morotai

"The P-61 chased two Japanese bogeys for four minutes, forcing them to jettison their bombs into the sea."

First Lieutenant John S.Guzey
Officer, 419th NFS

Allied forces continued to strengthen their position at Sansapor, on the Vogelkop Peninsula of western New Guinea, with the objective of the Philippines only 650 miles away. General Douglas MacArthur now planned an assault on Morotai Island, located approximately 20 miles off the northern coast of Halmahera Island. The assault would be coordinated with the invasion of Palau by Admiral Chester Nimitz's forces in the Central Pacific. Morotai is located halfway between the Vogelkop Peninsula and the Philippine island of Mindanao, while the Palau group is situated at the western end of the Caroline Islands, approximately 350 miles from Mindanao. Morotai is a circular island like Noemfoor. However, it is twice as large as the latter island and the interior has peaks that reach 3,500 feet. Both Morotai and Palau could be utilized to provide ground based support for MacArthur's upcoming invasion of the Philippines.

Intelligence sources indicated that Japanese forces were fewer in number on Morotai than on Halmahera. Plans called for the Allies to secure Morotai and bypass the much larger Japanese garrison on Halmahera. The Allies could then improve and expand the airfields on the island and provide support for MacArthur's recovery of the Philippines. Aircraft from Morotai could fly directly to the southern Philippines in order to help establish Allied air supremacy. The island's ideal location permitted Allied forces to also keep an eye on Japanese

Left to right: Second Lieutenant Lane Thompson and Captain Don Dessert. Thompson flew as the radar observer for Dessert dating back to night fighter training in Florida. Dessert and Thompson were together when the former struck a power line near Fort Lauderdale, removing two feet from his wing. *Courtesy of Don Dessert.*

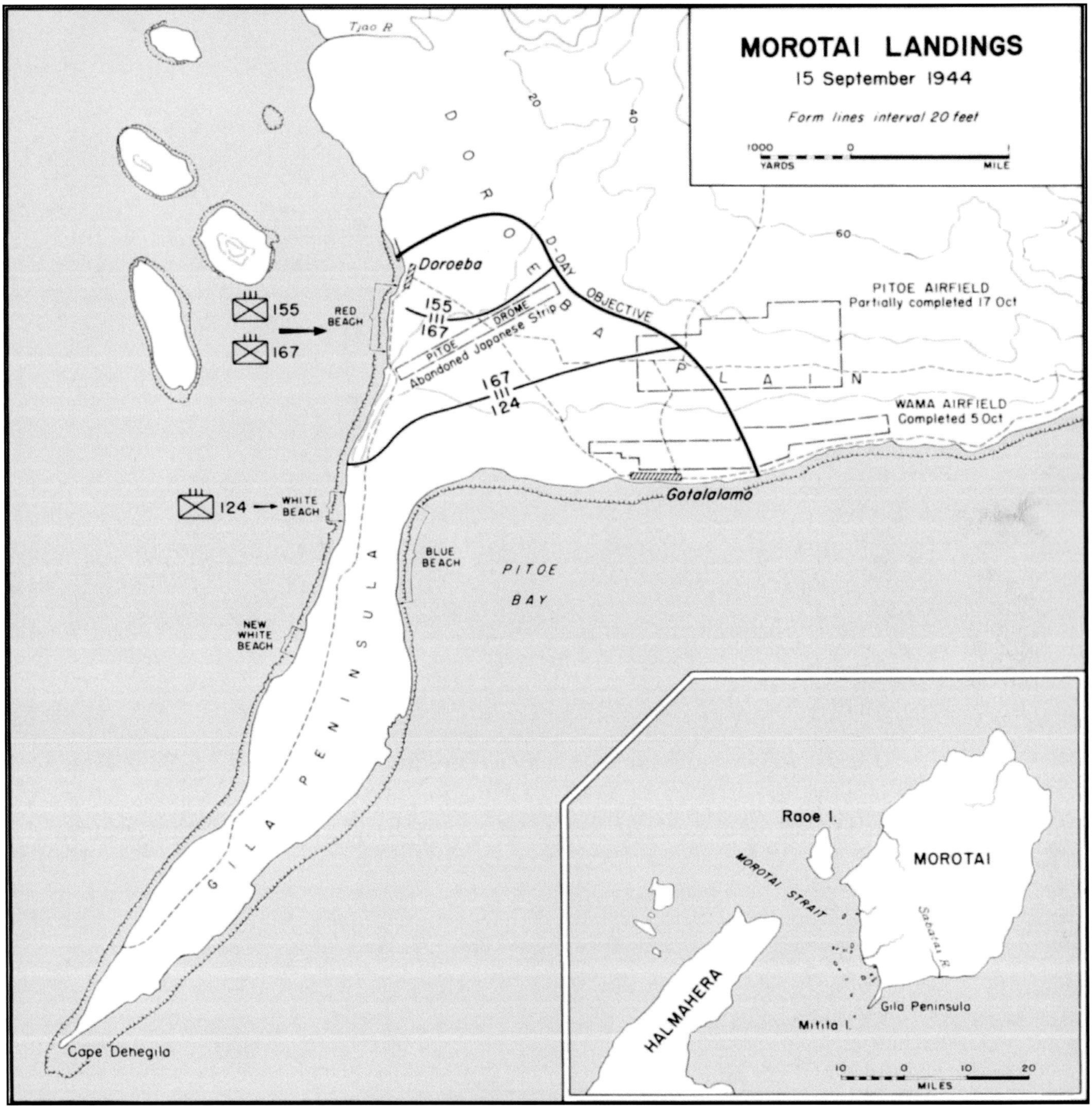

Morotai. *Courtesy of the Center for Military History.*

counter movements eastward from the northeast islands of the Dutch East Indies. Three regiments of the United States 37th Infantry Division landed at Morotai on September 15, 1944, and established a perimeter around the two airfields on the island.[1]

Japanese intruders conducted 82 night raids on Morotai between September 15, 1944, and February 1, 1945. Many of these raids involved sorties by multiple aircraft on the same night. The frequency of Japanese raids increased around the waxing and waning of a full moon. Before the unit's arrival on Morotai, the Northrop P-61 Black Widows of the 419th NFS staged through the island from their bases on Middleburg and Noemfoor in order to provide fighter coverage over the Allied fleet gathered for the invasion of the Philippines. In late September, First Lieutenant Don Dessert flew out of Morotai in support of the fleet supporting General Douglas

P-61 Black Widow 42-5531 #331. Personal transport of BG Earl Barnes, commander of 13th Fighter Command and later the 13th Air Force. Barnes selected this P-61 Black Widow, originally bound for the 419th NFS, as his personal transport aircraft. Maintenance crews removed the radar and installed a long range fuel tank. *Courtesy of Bob Fowble.*

MacArthur's landing at Leyte. Dessert referred to the large gathering of naval vessels as an "unbelievable sight" and noted he could see vessels "from horizon to horizon." Dessert became bored as he flew cover over the fleet. Spotting several naval aircraft in formation to land on an aircraft carrier, Dessert maneuvered his Black Widow to join them. It should be noted that carrier aircraft are modified with landing hooks that permit them to snag cables and thus stop quickly on the short aircraft carrier decks. The Black Widow naturally was not designed to land on an aircraft carrier. Dessert thought maneuvering into the landing pattern and appearing as though he planned to land on a carrier would be a good joke on the Navy. He proclaimed:

"I did get into the traffic pattern with the recovering Navy fighters landing on the carrier one evening. I put my [landing] gear down and made a pass over the carrier deck. I heard a radio message from the carrier stating 'we will shoot you down if you do that again.' I didn't!"[2]

Flight Officer Sanford "Sandy" Bauman, a Radar Observer (R/O), added details to the tale as related to him by Dessert's R/O (Jim Allen) on the mission:

"He [Dessert] flew a good base leg, turned on final, lowered his gear and flaps, and then all hell broke loose. The Navy had no sense of humor. Jim Allen, his R/O, said Dessert pulled vapor trails and many 'G's getting out of there."

Another view of P-61 Black Widow 42-5531 #331, the personal transport of BG Earl Barnes, commander of 13th Fighter Command and later the 13th Air Force. *Courtesy of Don Dessert.*

Allen himself noted years later that he remembered watching the "carrier landing officer with his two flags running, jumping, and frantically waving us off."[3]

While most of the Japanese intruder aircraft inflicted little damage to the Allied forces on Morotai, an attack on November 22, 1944, resulted in two deaths, 15 wounded, 15 destroyed aircraft, and eight damaged aircraft. Five days later, Detachment A of the 419th NFS arrived and worked with its sister unit from Florida, the 418th NFS of the 5th Air Force, until December 20, 1944. When the 418th NFS transferred from the island, Detachment A wove its spider's web along the southern approaches to Morotai to catch Japanese intruder aircraft.

The 419th NFS found defending Morotai from Japanese night intruders to be a challenge. Japanese forces still maintained their airfields on Halmahera despite the constant poundings they received from American aircraft. Most of the night intruders over Morotai originated from Ceram or the Celebes, in the Dutch East Indies. Celebes is the largest island between Borneo and New Guinea, while Ceram is a smaller island located off the southwest coast of New Guinea. Japanese crews piloted their aircraft from these islands to their airfields on Halmahera. The crews departed Halmahera during the night or early morning for the flight north to Morotai. Most intruders approached the American airfields from the south, and for this reason Detachment A developed a pattern of patrolling approximately 25 miles south of Morotai in the attempt to place themselves between the airfields and incoming Japanese aircraft. The American night fighters often had very little time to intercept the Japanese intruders before they entered the Morotai anti-aircraft (AA) gun circle.

Detachment A also had to contend with what was known as the "Halmahera Permanent Echoes." These were areas where Japanese planes could effectively hide from American radar. A 3,000+ foot ridge line ran across Morotai. Similar terrain could be found across the strait on Halmahera. Japanese pilots learned that they could maneuver into areas adjacent to these ridges and have their radar signatures masked by them. These became good areas to hide from pursuing P-61s or to infiltrate into the area unseen by Allied ground-based radar systems.

Japanese intruders hit Morotai the first night of Detachment A's arrival and continued for the next two nights. Second Lieutenant John S. Guzey, the 419th NFS assistant intelligence officer, commented:

"The ferocity of the raids and damage on the island made the construction of good fox holes an extreme necessity. However, the morale improved considerably over the previous month, for again, the men realized they were in combat and that they had an opportunity to show the results of their long period of extensive training."

On December 6, 1944, the first Japanese intruder felt the bite of a 419th NFS Black Widow over Morotai.

First Lieutenant Robert "Bob" Winslow, also known as "Snake" due to being tall and very thin, and his crew, Flight Officer Anthony Grotzinger (R/O) and

419th NFS Engineering Section led by First Lieutenant Michael Burlingham. The Engineering section performed a variety of maintenance services to the P-61 Black Widows to ensure the aircraft remained in operational status. The section put in many long hours evaluating and solving special problems with the P-61s, including the failure of the electrical system in the upper machinegun turret. The failure of the upper turret negatively impacted at least two P-61 intercept missions against Japanese aircraft. *Courtesy of AFHRA.*

Sergeant Mark Ridenour (gunner), conducted the early morning patrol from Morotai on 6 December. Winslow, piloting his P-61 at 12,000 feet and approximately 15 miles southwest of Wama Air Strip, received a radio transmission from the Ground Control Interception (GCI) operator at 4:00 AM warning of multiple bogeys approaching Morotai. The controller directed Winslow to orbit at his current location until a clear direction of the unidentified aircraft could be ascertained. Within minutes the controller vectored the P-61 crew on a 230 degree heading to intercept one of the incoming aircraft. Grotzinger established a local radar fix on the bogey at five miles and directed Winslow from that point. The Black Widow dropped to 8,000 feet and gave chase at 300 miles per hour. Winslow throttled back as he approached the approximate location of the unknown aircraft and made a visual sighting at 1,500

Left to right: Flight Officer Tony Grotzinger, First Lieutenant Herman Manke, and Second Lieutenant Bob Winslow. A P-61 Black Widow crew, including Winslow (pilot), Grotzinger (radar observer), and Sergeant Mark Ridenour (gunner), downed a Japanese Mitsubishi G4M Betty bomber in December 1944. *Courtesy of Don Dessert.*

feet. He slowly brought his P-61 to within 600 feet of the bogey and positively identified it as a Japanese Mitsubishi G4M Betty bomber. Winslow recalled:

"I ordered my gunner to fire, but he immediately notified me that he was unable to do so, as the turret would not operate. I then fired a short burst from my cannon and the Betty returned fire from her tail position."

The Japanese bomber pilot commenced a series of violent maneuvers to lose the pursuing Black Widow. Winslow continued to fire a series of short bursts of the 20 mm cannons and the gunner on board the Betty fired four or five bursts from his position in return. Winslow added:

"Finally I got within 200 or 300 feet and directly behind her and fired a long burst until I was out of ammunition. I believe I hit her at this time because I feel it was impossible to miss her at that range. She then slid off rapidly to the starboard and down, and I believed her to be out of control. At no time did I see her on fire."

Following the chase, the Japanese plane appeared to go down approximately 50 miles southwest of Wama Air Field and only 15 miles from the Japanese Kaoe Airdrome.

Winslow did not make any attempt to search for the Betty because he was out of ammunition and so close to the Japanese base. He returned to Wama Airstrip and landed at 4:35 AM. An inspection of the P-61 revealed that the Japanese gunner had been fairly accurate. Winslow added:

"I did not realize my plane was damaged until an examination after landing, which disclosed that the main spar and spoiler on the left wing was severely damaged by six 7.7 mm machinegun projectiles."[4]

Mitsubishi G4M Betty bomber. First Lieutenant Bob Winslow and his crew, Flight Officer Anthony Grotzinger (radar observer) and Sergeant Mark Ridenour (gunner), received official credit for a "probable kill" against a Betty bomber near Morotai in December 1944. The upper machine gun turret of the P-61 Black Widow jammed during the engagement, forcing Winslow to rely only on the 20 mm cannon in their fixed positions beneath the aircraft. The only way to aim fixed guns is to turn the entire aircraft. Winslow expended all 590 rounds of 20 mm ammunition that had been loaded into his Black Widow. "Finally I got within 200 or 300 feet and directly behind her and fired a long burst until I was out of ammunition. I believe I hit her at this time because I feel it was impossible to miss her at that range. She then slid off rapidly to the starboard and down and I believed her to be out of control. At no time did I see her on fire." *Courtesy of Bob Fowble.*

During the engagement, Winslow expended all 590 rounds of 20 mm ammunition that had been loaded into his Black Widow. It was much more difficult to bring down an enemy aircraft with just the 20 mm cannon of a P-61 because these guns were fixed. A pilot had to alter the position of the entire aircraft in order to aim them. The failure of the machine gun turret proved unfortunate during the engagement. The gunner should have been able to rotate the turret of a P-61 in order to obtain the best firing position when pursuing an enemy aircraft. Winslow and his crew received official credit for a "probable" and not a "kill" because the downing of the Betty could not be confirmed. The bomber went down very close to a Japanese airfield, prohibiting other Allied aircraft from confirming its destruction. In addition, the P-61 crew was not able to witness the crash. While the crew watched the Betty fall, apparently out of control after a near point-

Nose art on a 419th NFS P-61 Black Widow. *Courtesy of John Graham.*

419th NFS P-61 Black Widow 42-5497 #322 *Dawn. Courtesy of John Graham.*

blank range attack, no one actually saw it hit the ground. While the military would only grant a "probable" to the crew, Detachment A personnel were confident that they had brought down the Japanese intruder.

On the afternoon of 7 December a P-61 with First Lieutenant Bob Oates at the controls, Second Lieutenant Paul Ferguson (R/O), and Sergeant Alolpho Falcon (gunner) threw a propeller blade during takeoff from Morotai. The blade of the right propeller damaged the side of the crew nacelle and started an engine fire that damaged the power plant, ducts, and inter-coolers leading to the engine. The crew was not injured in the accident. Sergeant Harold Young, a mechanic, remembered one propeller blade looked like it had been "sawed off." "After that accident we pulled the propellers after every 25 hours of flight time to have them magnafluxed." Both 419th NFS detachments stood down while maintenance personnel checked the blades on every P-61 to ensure the accident would not be repeated.[5]

First Lieutenant James Schroth encountered a Japanese intruder three nights later on December 10, 1944. Schroth and his crew, newly promoted Second Lieutenant Fred James (R/O) and Sergeant Ridenour (gunner), were patrolling at an altitude of 12,000 feet 25 miles southwest of Wama Airstrip. At 3:00 AM the GCI controller radioed to inform the crew that multiple bogeys were in the area. James acquired a radar fix on four separate aircraft within minutes and directed Schroth to one of the targets flying five miles away from the P-61. The pilot brought his Black Widow within 2,000 feet of the unidentified aircraft but had to break contact as they entered the AA circle. The GCI controller directed the P-61 to orbit. Fifteen minutes later the controller vectored Schroth to a target approximately 25 miles south of Wama Airstrip. James picked up the bogey with his SCR 720 radar system when the P-61 pulled within four miles of the other aircraft. Schroth approached his opponent at 270 miles per hour at an altitude of 8,000

419th NFS Armament Section led by First Lieutenant Clarence (Tom) Erb. The Armament section serviced both the .50 caliber machine guns and 20mm cannon mounted on the P-61 Black Widows. Both types of weapons required regular cleaning, oiling, and preventive maintenance in the tropical climate to ensure they would function when needed in combat. Prior to the arrival of the P-61s, the 419th NFS salvaged bomb racks from old B-25s on Guadalcanal and utilized them to modify the P-70 Nighthawks inherited from Detachment B, 6th NFS. This permitted crews to fly the P-70s on night intruder missions over Japanese held areas. *Courtesy of AFHRA.*

Staff Sergeant John Graham in a rare view of a P-61 Black Widow gunner at his station. The photo is taken across the top of the pilot's cockpit, showing the slightly elevated position of the gunner. Note the optical gunsight and hand controls being demonstrated by Graham. Above and behind Graham is the P-61's upper turret housing four .50 caliber machine guns (although only two can be seen in this photo). The optical gunsight and hand controls electrically turned and aimed the gun turret. If an enemy aircraft maneuvered behind and out of the view of the gunner, he could electrically pass control of the gun turret to the radar observer in the rear of the aircraft. The radar observer had his own optical gunsight and hand controls as well as a 180 degree view toward the rear of the P-61 from his compartment. *Courtesy of John Graham.*

419th NFS Radar Section. First Lieutenant James Irwin assumed the title of squadron radar officer. The personnel assigned to the section performed the critical job of maintaining the temperamental SCR-720 radar system on board the P-61 Black Widows. The SCR-720 proved to be a major power drain on the P-61s and often failed at the worst possible times during missions against unknown bogeys. At other times, the system provided excellent information on bogeys at distances that sometimes exceeded five miles distance from the P-61. *Courtesy of AFHRA.*

feet. The target initiated a violent evasive maneuver when the P-61 moved within 300 feet of him. The bogey disappeared from the Black Widow's radar screen and James was not able to re-acquire him. The disappointed crew landed on Morotai at 4:15 AM.[6]

Staff Sergeant John Graham, a gunner, remembered riding shotgun one night with Schroth. Crews normally flew two patrols when on night intruder protection. They would fly an early evening mission, land and get some rest while the Black Widow was fueled and serviced, and then fly an early morning patrol. Schroth and his crew were returning to Morotai around 4:30 AM one morning, and the pilot requested that the airstrip landing lights be turned on as he approached. Consolidated B-24 Liberator bombers preparing for a mission to hit the oil facilities on Borneo often departed Morotai at that time of morning, and approximately 15 were lined up on a taxiway with their red and green night lights shining as they awaited permission to depart. Someone failed to turn on the airfield landing lights, and as Schroth approached the strip, he could see only the long line of red and green lights. Graham remembered Schroth telling his crew, "Those landing lights sure look strange." At the last moment the pilot decided to abort the landing and pulled up, buzzing the entire line of parked B-24s, fully fueled, loaded with bombs, and preparing to take

419th NFS Communications Section. The tropical climate of the Southwest Pacific played havoc on the communications equipment installed in the P-61 Black Widows, and this section worked diligently at all hours to ensure each pilot and radar observer team had what they needed to coordinate with the Ground Control Intercept (GCI) team and air traffic control team on the ground. Communications, like the radar system, could cease functioning at the worst possible times. During August 1944, First Lieutenant Herman Manke scrambled from Noemfoor in pursuit of an unidentified bogey north of the island. A problem with the onboard communications equipment nearly brought down his P-61 to friendly fire that evening. *Courtesy of AFHRA.*

off for their mission. Graham noted, "Imagine what would have happened if we had hit those bombers!" A startled Schroth chewed out the control tower personnel and they turned on the landing lights for a successful second approach.

Graham remembered another mission with Schroth. The GCI controller notified the Black Widow crew of an incoming bogey. Schroth and his crew were very excited and went after the bogey. The R/O made a local radar acquisition of the unidentified aircraft and Schroth moved in closer for a visual. As they approached the bogey, the crew noticed they trailed a North American B-25 Mitchell bomber that apparently did not have its IFF (Identification Friend or Foe) working properly. Without an IFF signal, the GCI controller would report the bird as an unidentified aircraft and direct an orbiting P-61 to investigate. Schroth was so upset that the bogey was a B-25 and not a Japanese plane he flew around the American bomber. Graham humorously recalled:

"The B-25 gunners were so terrified by the sudden encounter and concerned of being shot down they leaned out of their waist gun windows and shined flashlights at the star painted on the fuselage."[7]

A radio communications failure cost Detachment A a kill in mid-December. On 17 December Second Lieutenant David Amerman sat on ground alert at Wama Fighter Strip with his crew, Second Lieutenant Warren Bakewell (R/O) and Sergeant Adolfo Falcon (gunner). At 3:00 AM the GCI controller relayed the presence of a bogey approaching Morotai. Amerman and his crew scrambled their Black Widow and climbed to 14,000 feet in accordance with the controller's instructions. A warning from the CGI controller notified the crew that the bogey had entered the AA circle. Amerman waited approximately 15 miles southeast of Wama to see if the unidentified aircraft would emerge from the AA barrage. The GCI controller called at 3:15 AM to vector Amerman on a 90 degree course to intercept the bogey as it flew from the AA circle. Bakewell acquired a radar fix on the SCR 720 and directed the pilot for the final leg of the pursuit. The enemy aircraft flew at 12,000 feet only five miles from the P-61. Amerman increased his speed and quickly closed the distance to two miles. At that point the R/O reported the bogey had dropped off his radar screen.

The crew continued to search for the Japanese aircraft until a few minutes later, when the GCI controller reported the enemy aircraft had re-entered the AA circle. Amerman orbited his bird waiting for further instructions. The GCI controller presented a new vector for the P-61 at 3:30 AM. Amerman initiated a new pursuit, and Bakewell made a second radar acquisition of the Japanese aircraft at a distance of six miles. The R/O continued to direct the pilot until the Black Widow flew within 700 yards of the bogey. The crew made a visual identification of the aircraft as a Japanese Betty bomber flying 20 degrees above and 10 degrees to the port side of the P-61. Both planes were at 7,000 feet and flying between 150 and 170 miles per hour. Although Amerman lost his visual sighting of the Betty Falcon could still see it. Amerman commented:

"...I ordered him [Falcon] to fire. He fired two bursts [of the machine guns] with undetermined results. The gunner then observed bogey to peel off to the right."

The gunner had fired 150 rounds of .50 caliber bullets at the enemy bomber. Amerman tried to follow the Betty but was still not able to see it. Falcon attempted to inform the pilot of the Betty's location, but at that moment the Black Widow suffered a communications failure. The gunner and pilot could not communicate. The GCI controller called with a new vector to intercept the escaping Betty. Bakewell acquired a new radar fix on the Betty at 3:44 AM and the crew re-established their pursuit. The R/O guided the pilot to a position behind the Betty bomber. The pair of planes were now flying at 4,000 feet, and the Betty dropped even lower in an attempt to hide from the SCR 720 radar. The ground clutter from Halmaheras Island masked the Betty's

419th NFS P-61 Black Widow 42-5508 #313 *Uncle Sugar Able* (for U.S.A.). *Courtesy of John Graham.*

419th NFS P-61 Black Widow 42-39450 #329. *Courtesy of John Graham.*

Captain Al Lukas and unknown Royal Australian Air Force officer. Fighter pilots with the Royal Australian Air Force expressed dismay at being left behind on Morotai for ground attack missions against bypassed Japanese garrisons as American squadrons moved forward to the Philippines. Many senior Australian pilots presented their resignations in April 1945, prompting an investigation into their complaints. *Courtesy of Don Dessert.*

return signature on the radar scope and Bakewell lost the bogey. The crew continued their patrol for 15 minutes waiting to hear if the GCI controller could re-acquire a fix. Amerman landed at 4:40 AM.[8]

The 418th NFS (5th Air Force) departed Morotai on 20 December and Detachment A of the 419th NFS (13th Air Force) assumed sole responsibility for night fighter coverage over and around the island. Detachment personnel believed their P-61s were too few in number to provide adequate nocturnal protection of Morotai. The 419th NFS Headquarters Detachment on Middleburg dispatched two Black Widows and their crews the next day following a request for assistance. Additional aircrews continued to transfer from Middleburg throughout the rest of December while First Lieutenant Winslow, First Lieutenant Herman Manke, Flight Officer Grotzinger, and Flight Officer Sandy Bauman flew to Middleburg to await rotation back to the United States. That same night, a P-61 crew attempted an intercept of an unknown aircraft but the bogey remained too low for an accurate radar reading due to feedback on the system from ground clutter.

Second Lieutenant Clarence "Clem" Green, an R/O, noted an embarrassing intercept mission on 24 December. The Detachment flew a night patrol in cooperation with an Australian unit. While one P-61 patrolled at 18,000 feet and a second Black Widow at 12,000 feet utilizing radar, a Royal Australian Air Force (RAAF) Supermarine Spitfire fighter flew at 25,000 feet. The single seat British manufactured Spitfire did not have its own radar equipment, but rather waited for searchlights to lock onto the target. The Spitfire pilot would then confirm the type of aircraft captured in the searchlight's beams and attempt to destroy it if proven to be Japanese. This was the same procedure employed by the 419th NFS with the unit's Lockheed P-38 Lightning and could be effective, as demonstrated by First Lieutenant Richard Stewart's kill of a Japanese Betty bomber over Middleburg. Under the conditions of moonlit, cloudless skies, a Spitfire pilot could identify a Japanese aircraft and shoot it down without the benefits of radar or a searchlight. While this feat was difficult, it was certainly not impossible, as demonstrated by numerous night fighter kills by European pilots early in World War II.

Green noted that he acquired a "blip" on his radar screen, indicating the presence of a bogey in the area. The unknown plane's pilot knew that he was being pursued and opted to depart the area. Green observed the "blip" moving up and off his screen as the bogey increased speed. Green directed his pilot, First Lieutenant John MacLeod, in the chase, but the crew soon gave up when

419th NFS Intelligence Section reading table. Intelligence clerk Sergeant Ralph Dunavant sits nearest to the camera at the right side of the table. Note the 13th Air Force patch on his sleeve. The unit maintained manuals and magazines in this area for everyone to read. A large radio can be seen at the back of the room. *Courtesy of Leslee Lukas Mays.*

they lost and could not re-acquire radar contact with the other aircraft. Green added:

"We gave up the chase and returned to base to be greeted by an Aussie [Australian] airman who asked us if we'd downed the bogey. When we replied 'no,' it was then that we first learned one had been downed by a Spitfire. Very humiliating."[9]

Six days later Sergeant Falcon flew as the gunner for First Lieutenant Williamson (pilot) and Second Lieutenant Willard Riley (R/O). Williamson and his crew were patrolling approximately 15 miles southwest of Wama Fighter Strip on 27 December. Visibility this night was very poor. The GCI controller reported the existence of a bogey approaching Morotai from the south. The controller informed Williamson to alter his course 180 degrees and drop to an altitude of 11,000 feet. Somehow the Black Widow crew missed the bogey as it flew past them. The GCI controller called for a course correction to 350 degrees to catch the unidentified aircraft. Williamson complied with the instructions and initiated a pursuit while dropping his altitude to 9,000 feet. Riley acquired a radar fix of the bogey at a distance of one mile and the crew prepared to make a kill. Williamson moved into a position within 500 feet of the enemy aircraft and called for a visual scan from the crew members. The pilot noted:

"...the target made a hard dive to port and was lost. Just before diving, the target was seen momentarily by the gunner who did not identify same, but did see a plane with green and red navigation lights on."

At 4:58 AM the GCI controller called and vectored Williamson to the southeast. The bogey never entered the AA circle around Morotai and the P-61 rose to 11,000 feet in pursuit. Riley re-established radar contact and informed the pilot that the bogey was now five miles away from the P-61. The R/O lost the radar fix

Major Joseph Shulmistras, who replaced Major Andrew Bing on July 1, 1944, as commander of the 419th NFS. Bing, a veteran pilot of the North African campaign, commanded the 419th NFS for only two weeks before being transferred to 13th Fighter Command. *Courtesy of Don Dessert.*

and then re-acquired it at a range of just over two miles. As the Black Widow approached closer the crew lost its radar fix and the bogey. Williamson landed at 5:45 AM after making additional attempts to acquire the bogey's position with the SCR 720.

First Lieutenant Amerman tackled a Japanese bomber crew that decided to put up a fight the next night. On 28 December Amerman and his crew of Second Lieutenant Bakewell (R/O) and Staff Sergeant George Russell (gunner) were patrolling 25 miles south of Wama Fighter Strip. The GCI controller reported the presence of multiple unidentified aircraft approaching Morotai. The controller vectored Amerman onto a 160 degree course at 19,000 feet. At seven miles Bakewell established radar contact with one bogey. The pilot gave chase and approached to within 3,000 feet of the target. Numerous return signatures appeared on the SCR 720 radar scope indicating the enemy aircraft was utilizing "window" (foil strips dumped from the plane) to produce false images on radar. Williamson remembered, "The bogey evidently could see us very well." Russell made a visual

419th NFS mess hall and staff. The talents of the mess personnel were limited by the ingredients they were provided. On Guadalcanal, mutton was the only fresh meat available to the mess section. Captain Dick Stewart remembered, "You could smell the mutton cooking in our alert shack which was half a mile away from the mess hall. It was cooked in a single large pot and had a couple inches of grease floating on top." The officers frequently collected money and dispatched a plane to acquire fresh food. The squadron C-47 delivered fresh fruit and vegetables. Captain Don Dessert mentioned that the unit "loaned" a P-38 to the Marines in the Philippines in exchange for a walk-in refrigerator, after which Dessert remarked, "We had the best mess hall in the Pacific." *Courtesy of Don Dessert.*

P-61 Black Widow maintenance. Crushed coral proved to be a significant problem facing the maintenance crews of the 419th NFS. The highly abrasive substance scratched plexiglas and could choke engines, leading to at least one fatal take off attempt. *Courtesy of Dot McConnell.*

sighting at 1,000 feet. Although he could not positively identify the type of aircraft, Russell noted that it was two-engine, low winged, and painted a dark color. The enemy bomber turned into the moon that shone that night and continued to the starboard in an attempt to get behind the Black Widow pursuing it. Williamson continued:

"As the bogey turned, he fired a short burst at us from a tail gun, or possibly a top or belly turret. He then turned hard starboard and we followed him through three complete revolutions, in which we held the bogey at 80 to 90 degrees starboard. He fired another burst, probably from a tail gun or turret."

The Black Widow pilot commented later about his confusion over the source of the bomber's defensive fire. Although he chased the enemy bomber, he felt that the gun fire could not have come from a tail position. "I don't see how it could have been a tail gun, unless the gun had a full 180 degree coverage to the rear." The enemy plane banked sharply after firing the second time and disappeared from the radar screen. The GCI controller warned Amerman that he was approaching the Morotai AA circle so the pilot throttled back and discontinued the pursuit. The controller then directed the Black Widow to a second bogey approaching the island. Bakewell obtained a local radar fix within minutes of the warning. However, the enemy aircraft entered the AA circle before the Black Widow could catch it. Amerman orbited in the area waiting for additional instructions from the GCI controller. As both bogeys departed the AA circle, the GCI controller could not establish solid readings and both quickly dropped off the screen. The frustrated Black Widow crew landed at 5:35 AM.[10]

Line Chief Master Sergeant Bill McConnell standing in front of 419th NFS P-61 Black Widow 42-5515 #316 *Sleepy Time Gal*. At least three different night fighter squadrons had P-61s nicknamed *Sleepy Time Gal. Courtesy of Dot McConnell.*

Left to right: First Lieutenant Ralph LeVitt (pilot), Second Lieutenant Freddy Kahn (radar observer), and Staff Sergeant John Graham (gunner). The trio downed a Japanese Mitsubishi G4M Betty bomber during January 1945. LeVitt inched the P-61 to within 400 feet of the Japanese bomber without being spotted by its crew. He later reported, "I fired the 20's [cannon] and the gunner fired the 50's [machine gun]. The Japanese plane started burning after being hit by the first round. The fire lit up the whole ship...." Graham added, "I turned my guns on him and started firing and LeVitt fired his guns at him. The Japanese plane went down just spinning on fire." *Courtesy of John Graham.*

December 1944 proved to be an active month for Japanese intruders over Morotai. Japanese bombers flew over Morotai at least 20 of the 31 nights in the month, including periods without a full or partial moon. The largest Japanese raid involved the four aircraft; however, they flew individually across Morotai rather than as a single formation where their bombs, if dropped on target, could have made a much larger impact on American operations and morale. January would be a much lighter period for Japanese intruder activity, with raids on the 1st, 4th, 8th, 9th, and 11th of the month. The intruder on January 9, 1945, flew into a Black Widow's web and did not escape its sting.[11]

First Lieutenant Ralph LeVitt made one of the 419th NFS' most impressive kills on the night of January 9, 1945. Bagging a Japanese bomber that night resulted from close cooperation and calculation between the pilot, R/O, and GCI controller. LeVitt flew a regular patrol that night with Second Lieutenant Frederick "Freddy" Kahn (R/O) and Staff Sergeant Graham (gunner). The crew were south of Wama Air Strip when the GCI controller, First Lieutenant Charles Bramlett, radioed the presence of multiple bogeys approaching Morotai. Bramlett vectored LeVitt to a 270 degree intercept course and directed the pilot to climb from 17,000 to 20,000 feet. The controller continued to adjust the vector as the Black Widow approached the unidentified targets. Bramlett realized that the Japanese planes were heading for one of the "Halmahera permanent echoes" where their radar signatures would be masked from the GCI and the P-61 SCR 720 radar system. In response, he directed LeVitt to a location where he thought the Japanese planes would enter the area based upon their compass heading. The tactic was a success, and at 2:45 AM Kahn acquired a radar fix with the P-61's SCR 720 as the bogeys attempted to enter the "Halmahera permanent echoes." He informed LeVitt that the bogeys were 2 miles away from Black Widow. At this point the P-61 was approximately 25 miles south of Wama Air Strip. LeVitt overshot the target without ever making visual contact with the incoming aircraft. The pilot then zig-zagged his Black Widow to help Kahn re-acquire a radar fix on the bogeys. Kahn finally picked up a radar signal and informed LeVitt that a single target was now flying 1 ½ miles in front of them.

Radar Observer Second Lieutenant Freddy Kahn in his compartment at the rear of the P-61 Black Widow. The radar scope is located in front of the radar observer, who flew facing forward. Note the good view provided by the radar observer's compartment, permitting him to see 180 degrees behind the P-61. Kahn, along with First Lieutenant Ralph LeVitt (pilot) and Staff Sergeant John Graham (gunner), tracked and downed a Japanese Mitsubishi G4M Betty bomber. Kahn proved the value of the SCR-720 radar system onboard the P-61. Graham noted that he could not see the Japanese aircraft, even though Kahn kept telling him it was just off their starboard position. "All I could see was the black sky. Then one of the stars disappeared and I called out 'I can see him.'" *Courtesy of Don Dessert.*

LeVitt initiated a chase and flew to within 800 feet of the bogey's radar position in order to begin a visual search. Graham noted that he could not see the Japanese aircraft even though Kahn kept telling him it was just off their starboard position:

"All I could see was the black sky. Then one of the stars disappeared and I called out 'I can see him.' His silhouette covered the star, providing me with a reference point."

Both pilot and gunner spotted the Japanese aircraft flying above and slightly starboard of their position. LeVitt inched the P-61 to within 400 feet of the Japanese Betty bomber without being spotted by its crew. He reported:

"I fired the 20s (cannon) and the gunner fired the 50s (machine gun). The Japanese plane started burning after being hit by the first round. The fire lit up the whole ship…." Graham added, "I turned my guns on him and started firing and LeVitt fired his guns at him. The Japanese plane went down just spinning on fire."

The Black Widow crew peeled off at a range of 100 feet from the Betty in order to avoid a collision and watched as the bomber spun downward and appeared to explode just prior to hitting the water. LeVitt utilized 200 rounds of 20 mm cannon ammunition and Graham expended approximately 300 rounds of .50 caliber ammunition making the kill.

Kahn picked up a second contact on his radar screen after the crew knocked off the Betty. However, the bogey entered the Morotai AA circle, preventing the Black Widow from pursuing. At 2:53 AM the GCI controller radioed a new contact flying 12 miles away and directed the P-61 onto a 200 degree vector. The R/O acquired two returns on his scope, indicating the presence of multiple Japanese aircraft. Both bogeys were conducting wild evasive maneuvers in an attempt to lose their pursuer.

First Lieutenant Ralph LeVitt in the pilot's cockpit of the P-61 Black Widow as seen from the gunner's compartment. LeVitt and his crew, Second Lieutenant Freddy Kahn (radar observer) and Staff Sergeant John Graham (gunner), bagged a Japanese Mitsubishi G4M Betty bomber in January 1945. Immediately after that kill Kahn picked two additional bogeys on his SCR-720 radar system. LeVitt initiated a pursuit based on Kahn's vectoring. Both bogeys flew wild evasive maneuvers in an attempt to lose their pursuer. LeVitt chased both Japanese planes for four minutes, forcing them to harmlessly jettison their bomb loads as they withdrew from the area. While the P-61 crew did not bag a second Japanese plane that night, the Black Widow proved its worth by preventing the intruders from dropping their ordnance on American military personnel on the ground. *Courtesy of Don Dessert.*

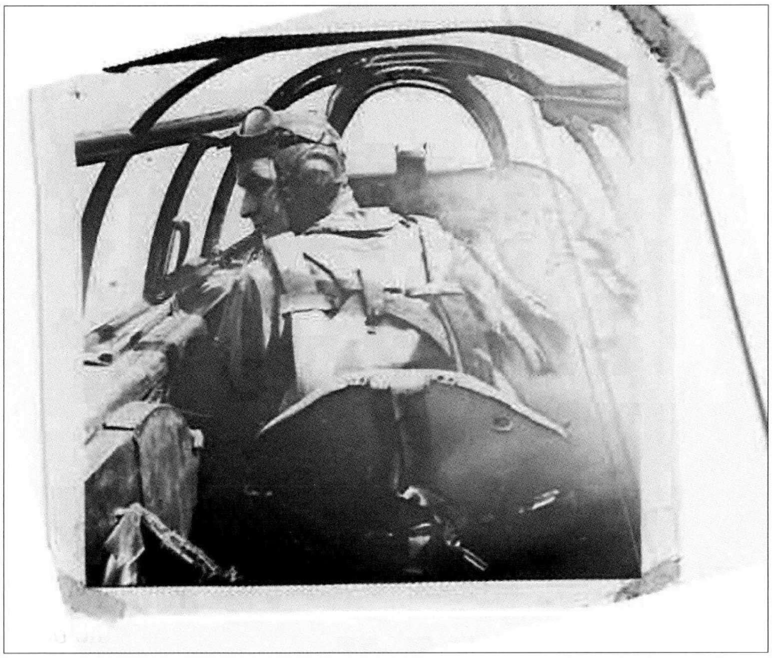

Another view of LeVitt in the pilot's cockpit of the P-61 Black Widow. The photo is taken from the gunner's compartment located behind the pilot. *Courtesy of John Graham.*

LeVitt chased both Japanese planes for four minutes, forcing them to jettison their bomb loads. The crew managed to get within 1,000 feet of one aircraft and made a fleeting visual contact before losing it. Kahn could not re-acquire a radar fix so the GCI controller vectored the P-61 toward another bogey leaving the area. Graham recalled "One of the best ways to get away was to dive and get close to the water so the radar couldn't pick you up." LeVitt dropped to 10,000 feet altitude and Kahn made momentary radar contact before losing the escaping plane as it dove. The controller then directed the P-61 to return to Wama airfield and land. After the engagement, the P-61 crew noted that credit for the successful kill should also be given to Bramlett's proactive thinking and tactics.[12]

Sergeant Jesse Gardner, an operations clerk, witnessed the kill:

"We had two alerts last night. The first was a cloud. [Thunderheads frequently gave false returns on early radar systems – author] The second was two Japanese planes. Lieutenant LeVitt with Kahn and Graham shot one down in flames into the ocean. I thought it was a flare at first."[13]

Second Lieutenant Green witnessed the destruction of the Betty bomber by LeVitt and his crew:

"I was standing next to Major Shulmistras [419th NFS commander] and could hear the motors of the two planes and the guns firing from LeVitt's aircraft. I watched the fireball descend out of the clear sky through a cloud, illuminating it from the inside, into the clear, and finally into

First Lieutenant Phil Rand leaning on an emitter of the P-61 Black Widow's SCR-720 radar system. During January 1945, Rand and his crew pursued a bogey identified by the Ground Control Intercept (GCI) station. Despite excellent communications and GCI-airborne coordination, the P-61 crew could not identify the bogey for an attempted shoot down. The GCI team provided excellent vectoring that brought the P-61 close enough for the radar observer (R/O) to acquire a signal on his SCR-720 radar system. First Lieutenant John Guzey, the assistant 419th NFS intelligence officer, blamed the lack of a kill on the poor visibility. "GCI Controller and R/O did a beautiful piece of intercepting. Visual was impossible due to weather and blackness. Estimated 100 feet to be the furthest distance possible to see. Had a solid bank of thunderheads southeast of base and cloud layers at various altitudes. Stars were even blacked out." *Courtesy of Don Dessert.*

the ocean just over the horizon. Ralph and Freddy were accorded a heroes' welcome upon their return to the squadron area."[14]

The next bogey to approach Morotai flew into the area on 11 January. First Lieutenant Phil Rand flew the regular patrol that early morning. The GCI controller directed Rand to climb to 20,000 feet to intercept multiple unidentified radar contacts. In a scenario experienced by so many 419th NFS pilots, the radio communications failed. Rand could not hear the GCI controller after reaching 19,000 feet. He continued to orbit in the area until his R/O established a local radar fix. The SCR 720 indicated that two bogeys approached Morotai. Rand picked one of the aircraft and the R/O directed him to the target. The enemy aircraft flew at 20,000 feet and an airspeed of 180 miles per hour. Rand crept to within 300 feet of the enemy aircraft. He noted later that visibility was very poor and he could not get a visual sighting of the aircraft. The pilot continued his chase for several minutes before opting to throttle back out of concern of ramming the bogey. The enemy aircraft turned sharply and the Black Widow's crew lost their radar fix. At 3:16 AM the R/O re-acquired the radar fix and closed to within 4,000 feet. At this point the GCI controller directed Rand to drop the chase due to his proximity to the AA circle. The gunners on Morotai had already begun to fire upon the bogey. Rand made a 180 degree turn to depart the area and then orbited to await the enemy aircraft. At 3:22 AM the R/O made a new radar fix, and the chase continued with a five mile separation between the two aircraft. The Japanese plane dove to an altitude of 14,000 feet and the Black Widow crew lost its radar fix during this maneuver. Rand elected to return to Morotai and landed at 4:45 AM. Communications and pilot-R/O coordination with the GCI controller could not have been better that morning. First Lieutenant Guzey blamed the lack of a kill on the visibility:

"GCI Controller and R/O did a beautiful piece of intercepting. Visual was impossible due to weather and blackness. Estimated 100 feet to be the furthest distance possible to see. Had a solid bank of thunderheads southeast of base and cloud layers at various altitudes. Stars were even blacked out."[15]

The Allies operated two airstrips on Morotai: Wama Airstrip for fighters and Pitoe, parallel to Wama, for bombers. The airstrips were approximately one-half mile from each other. At 4:00 AM on 12 January Sergeant Gardner was on the night shift for the 419th

419th NFS P-61 Black Widow 42-5495 #321. *Courtesy of John Graham.*

NFS at Wama Airstrip. He remembered watching fully fueled B-24 bombers taking off from Pitoe adjacent to his location:

"One of them had a tire blow out while taking off. He veered off the strip, hit an earthen embankment, and exploded as all three bombs he carried went off at once...There was a big fire and ammunition kept going off all the time."

Only one crewman survived the mishap.[16] Earlier in the 419th NFS' tour on Morotai the position of the two airfields may have saved the life of First Lieutenant Bob Oates one night during an alert scramble. First Lieutenant Oates was scrambled by the GCI controller to intercept an incoming bogey. The enemy aircraft arrived before Oates could get airborne to engage him. First Lieutenant John Radzieta remembered that as Oates was taking off on Wama Airstrip a Japanese fighter was flying parallel to him and strafing Pitoe Airstrip. If the Japanese intruder had chosen Wama Airstrip he may have been able to strafe Oates during the vulnerable time when engaged in a takeoff. The bogey managed to escape before Oates could get him.[17]

Second Lieutenant Lane Thompson. He flew as the radar observer (R/O) for Captain Don Dessert, who remembered Thompson as a well educated man with a tendency to sing while sitting in the rear R/O compartment of the P-61s overseas. *Courtesy of Don Dessert.*

The night after Gardner witnessed the B-24 accident Detachment A lost a P-61 during a landing attempt. First Lieutenant George White retuned to Wama Airstrip following a routine night patrol over Morotai. White's

419th NFS P-61 Black Widow 42-5495 #321. *Courtesy of Don Dessert.*

419th NFS P-61 Black Widow42-5495 #321. *Courtesy of John Graham.*

Left to right: Captain John Laffey and First Lieutenant Bob Oates. Laffey, a test pilot when serving with the Royal Air Force earlier in the war, could not pass an opportunity to shuttle a North American P-51 Mustang fighter to Leyte, in the Philippines, where his brother was assigned. Although someone pointed out that the plane appeared to have a slow fluid leak, Laffey brushed it off and went up in the bird. No one ever saw or heard from him again, and he is presumed to have died in a crash at sea. *Courtesy of Don Dessert.*

Captain Al Lukas developed a severe case of malaria while on leave and spent the next three months in a hospital in Australia. He recovered in time to join the 419th NFS in the Philippines prior to his end-of-tour rotation back to the United States. *Courtesy of Leslee Lukas Mays.*

crew included Second Lieutenant Burton Tracy (R/O) and Sergeant Alberto Murray (gunner). The strip was very wet, and upon landing White's Black Widow began skidding. The plane ran off the runway into the soft mud, collapsing the nose wheel assembly and severely damaging the nose of the plane. The crew was not injured, but the P-61 was relegated to the salvage heap.[18]

The Japanese monitored the Allied radio communications traffic at Morotai, and this proved troublesome on 20 January. First Lieutenant Kenneth Voy was flying to Morotai from Middleburg Island and asked for a homing transmition so he could find Wama Airstrip. The Japanese were listening to the communications and immediately began jamming the frequency in an attempt to prevent Voy from locating the base. The P-61 pilot switched to another channel, re-established contact with Wama Airstrip, and arrived safely.

At the end of January three 419th NFS aircrew members, Captain John Laffey (ex-Royal Canadian Air Force pilot), First Lieutenant John Blankenship (R/O for Captain Lukas), and Second Lieutenant Edward Kelly, departed for Middleburg to await their rotation orders

First Lieutenant Logan Southfield (seated at left) and Second Lieutenant Ben Goldstein (standing at far left) of the 550th NFS. Southfield piloted the P-61 Black Widow (42-39445) carrying Goldstein (his radar observer) and two other passengers on a routine hop from Hollandia, in New Guinea. The plane crashed on the slope of Mount Cyclops. Southfield recalled, "The aileron and spoiler systems froze making it impossible to turn away from the mountain. I therefore gave the engine full R.P.M., manifold pressure, and water injection. When I realized that the plane was about to crash, I warned the crew to brace themselves, ejected the hatch, and closed the switches. The plane crashed into the trees shedding a wing and the tail empennage." The Mid-Atlantic Air Museum recovered the P-61 during a 1990-1991 expedition and is restoring it to flying condition. *Courtesy of AFHRA.*

back to the United States. While awaiting orders, Laffey seized upon an opportunity to shuttle a North American P-51 Mustang fighter to Leyte, in the Philippines, where his brother was assigned. He had never flown this type of aircraft before and itched to take her up. Although someone pointed out that the plane appeared to have a slow fluid leak, Laffey brushed it off and went up in the bird. No one ever saw or heard from him again, and he is presumed to have died in a crash at sea. During this same period Captain Lukas developed a severe case of malaria on the way to rest leave in Australia and would spend the next three months in the hospital.

Most of the 419th NFS aircrews on Middleburg transferred to Morotai by the end of January while the Headquarters Detachment moved ahead to the Philippines, where it prepared for the arrival of the flight crews in March 1945. Two P-61s with their crews remained on Middleburg Island for night alerts while the newly arrived 550th NFS prepared to assume the night fighter mission over that island and Sansapor. The 550th NFS, under the command of Major Robert Tyler, arrived in the Pacific Theater at the end of December 1944 to become the third night fighter squadron to have served under the 13th Air Force during World War II. The squadron established its headquarters at Hollandia, on New Guinea, where the unit lost its first aircraft in theater on January 10, 1945. First Lieutenant Logan Southfield (pilot) departed Hollandia on a local hop carrying Second Lieutenant Ben Goldstein (R/O), Second Lieutenant Floyd Lovay (passenger from the 360th Troop Carrier Squadron), and Sergeant Ruth Hillman (passenger from the local Women's Army Corps detachment). Southfield buzzed the Hollandia airstrip and the local area before heading toward Cyclops Mountain at the edge of the base. The aircraft was seen climbing to fly over the mountain, but instead disappeared into the vegetation near the top. The P-61 did not re-emerge from the forest. Southfield recalled:

"The aileron and spoiler systems froze, making it impossible to turn away from the mountain. I therefore gave the engine full R.P.M., manifold pressure, and water injection. When I realized that the plane was about to crash I warned the crew to brace themselves, ejected the hatch, and closed the switches. The plane crashed into the trees, shedding a wing and the tail empennage."

The pilot administered first aid to Goldstein, who was the only person on the Black Widow to suffer an

Left to right: 419th NFS commander Major Joseph Shulmistras and First Lieutenant Don Gardner. Shulmistras commanded the 419th NFS from July 1, 1944, to January 30, 1945. *Courtesy of Don Dessert.*

injury. Southfield left the others at the crash site and climbed down the mountain to seek assistance. During the one hour descent the pilot slipped and fell numerous times, and in the process cracked a vertebra near his pelvic bone and chipped the bone in his heel. A search party rescued the other three at the crash site. Goldstein made a quick recovery. However, medical personnel placed Southfield into a cast and returned him to the United States at the end of February. The aircraft was a total loss, but the Mid-Atlantic Air Museum (Reading, Pennsylvania) recovered the remains nearly five decades later for full restoration.[19]

The last 419th NFS Black Widows transferred from Middleburg to Morotai on 3 February. With the arrival of all aircrews on Morotai the 419th NFS reorganized itself. Captain Dick Stewart assumed temporary command of the squadron on 1 February, and Major Shulmistras transferred to the 13th Fighter Command. Captain Don Dessert became the operations officer while Captain Bob Oates, First Lieutenant Bernard Knott, and First Lieutenant John McGhee assumed the role of flight leaders.

The 550th NFS moved its assets from Hollandia and Middleburg to Morotai beginning 14 February in preparation for replacing the 419th NFS as the primary night fighter unit on the island. February 1945 was a relatively quiet but rainy month for the 419th NFS. The aircrews flew only 110 combat missions, with 107 of these routine patrols without any contacts. One mission was a search operation, and the other two were night scrambles after the GCI controller warned of unidentified aircraft in the area. No radar contacts were acquired during these two missions and no enemy aircraft flew over Morotai. Unit personnel put the down time to good use and conducted a series of training classes. First Lieutenant David Guy, the GCI controller attached to the 419th NFS, received a Bronze Star medal on 17 February. He is the only person assigned or attached to the 419th NFS to receive a Bronze Star

Left to right: Major Joseph Shulmistras and Staff Sergeant John Radzieta standing in front of the 419th NFS' Douglas A-24 Banshee utilized for training and administrative hops. *Courtesy of John Radzieta.*

419th NFS medical section. The squadron was authorized its own medical doctor, dentist, and staff. *Courtesy of Don Dessert.*

Unknown 419th NFS maintenance NCO. *Courtesy of Dot McConnell.*

while with the unit. Guy was instrumental as the GCI controller during many of the 419th NFS kills, and many aircrew members praised the lieutenant for his expertise and skill in the challenging job.

The first two weeks of March proved to be relatively quiet and free of Japanese intruders. However, a rash of accidents struck the 419th NFS during this period. The squadron's Douglas A-24 Banshee was damaged in a taxiing accident. First Lieutenant Metzler was sitting in the cockpit when a taxiing RAAF Beaufort hit the A-24, damaging its wing and the propeller. In another incident, First Lieutenant Guy Scheetz was on patrol in his P-61 when the air filter in his engine caught fire. He made an emergency landing, blowing one tire and damaging another. First Lieutenant John MacLeod damaged the wing tips of his P-61 and that of another Black Widow when maneuvering his plane into a bunker. Despite the accidents, the men of the 419th NFS prepared for their next move. On 15 March the aircrews began shuttling their Black Widows to weave new webs in the Philippine Islands, where the ground crews from the Headquarters Detachment already awaited them after leapfrogging from Middleburg Island.

The 550th NFS did not remain much longer at Morotai. An advance element of the 550th NFS joined a detachment of the squadron at Tacloban, in the Philippines, on 13 March. On 18 March the main body of the 550th NFS discontinued operations on Morotai and prepared to move to the Philippines. By 7 April all elements of the 13th Air Force's night fighters had completely departed Morotai and were in the Philippines.

First Lieutenant Ralph Lillard, who flew as the radar observer for First Lieutenant Kenneth Voy. The Japanese attempted to confuse Voy and Lillard during a P-61 Black Widow flight from Middleburg Island to Morotai. The Japanese were listening to their communications and began jamming the frequency in an attempt to prevent Voy from locating the base. The P-61 pilot switched to another channel, re-established contact with Wama Air Strip, and arrived safely. *Courtesy of Don Dessert.*

7

Philippines

"The rear gunner of the [Japanese] Nick opened fire and tracers flashed by the Widow with no noticeable hits"

First Lieutenant Harold E. Michels
Pilot, 419th Night Fighter Squadron

As General Douglas MacArthur's forces moved throughout the Philippines the night fighters shifted with them. The 419th NFS provided air cover for the initial October 1944 invasion of the Philippines at Leyte from its base on Morotai. MacArthur landed on Luzon Island in January 1945. By March 1945 elements of the 419th NFS and 550th NFS transferred to the Philippines to provide night fighter support for continuing American operations in the archipelago as well as an Australian-led offensive on Borneo, the largest island of the Dutch East Indies (modern Indonesia). In the Philippines, the 419th NFS and 550th NFS intercepted Japanese night intruder aircraft; flew protective cover over Allied naval vessels and beachheads; located unknown ships at sea; attacked Japanese ground targets; and rescued lost Allied aircraft at night.

General MacArthur's initial invasion of the Japanese-held Philippines occurred on October 20, 1944, as four Army divisions of the Sixth Army landed on the island of Leyte. Lieutenant General Walter Krueger commanded the Sixth Army, while Lieutenant General George Kenney continued to lead the Far Eastern Air Forces and Brigadier General Earl Barnes headed the 13th Fighter Command of Brigadier General Paul Wurtsmith's 13th Air Force. It is interesting to note that the 419th NFS remained under the 13th Fighter Command while the 550th NFS was placed under the 13th Bomber Command. Vice Admiral Thomas Kinkaid, MacArthur's naval chief, commanded the naval forces, including amphibious assets, naval gunfire support, and close air support by planes based on escort carriers. MacArthur provided unified command over the three elements—air, land, and sea. While at sea, the Navy provided air cover for vessels associated with the Philippines landings. However, as soon as the vessels sailed within operating distances of land-based aircraft, the Far Eastern Air Forces of General Kenney shared this responsibility.

American landings in the southern and western Philippine Islands commenced in February 1945 under the code name series VICTOR with the 13th Air Force providing air support and the 419th NFS leading the effort in night fighter operations. They were soon joined by the 550th NFS. Flying cover in the area of the various invasion forces could be hazardous for pilots, and even more so for those of the 419th NFS, who were airborne at dusk, dawn, or during the night. Second Lieutenant Clifford Johnson, a 419th NFS pilot, recalled flying cover for naval vessels when he received a warning over the radio from the Fighter Control on board one of the ships. "The controller warned me to fly further out to sea since we might be in the path of the shells they were lobbing onto the coast." Johnson had forgotten that naval guns placed their projectiles into high arcs in order to achieve long distances to a target.[1] The Northrop P-61 Black Widow pilots contended with many hazards, including Japanese aircraft, darkness while airborne, and friendly

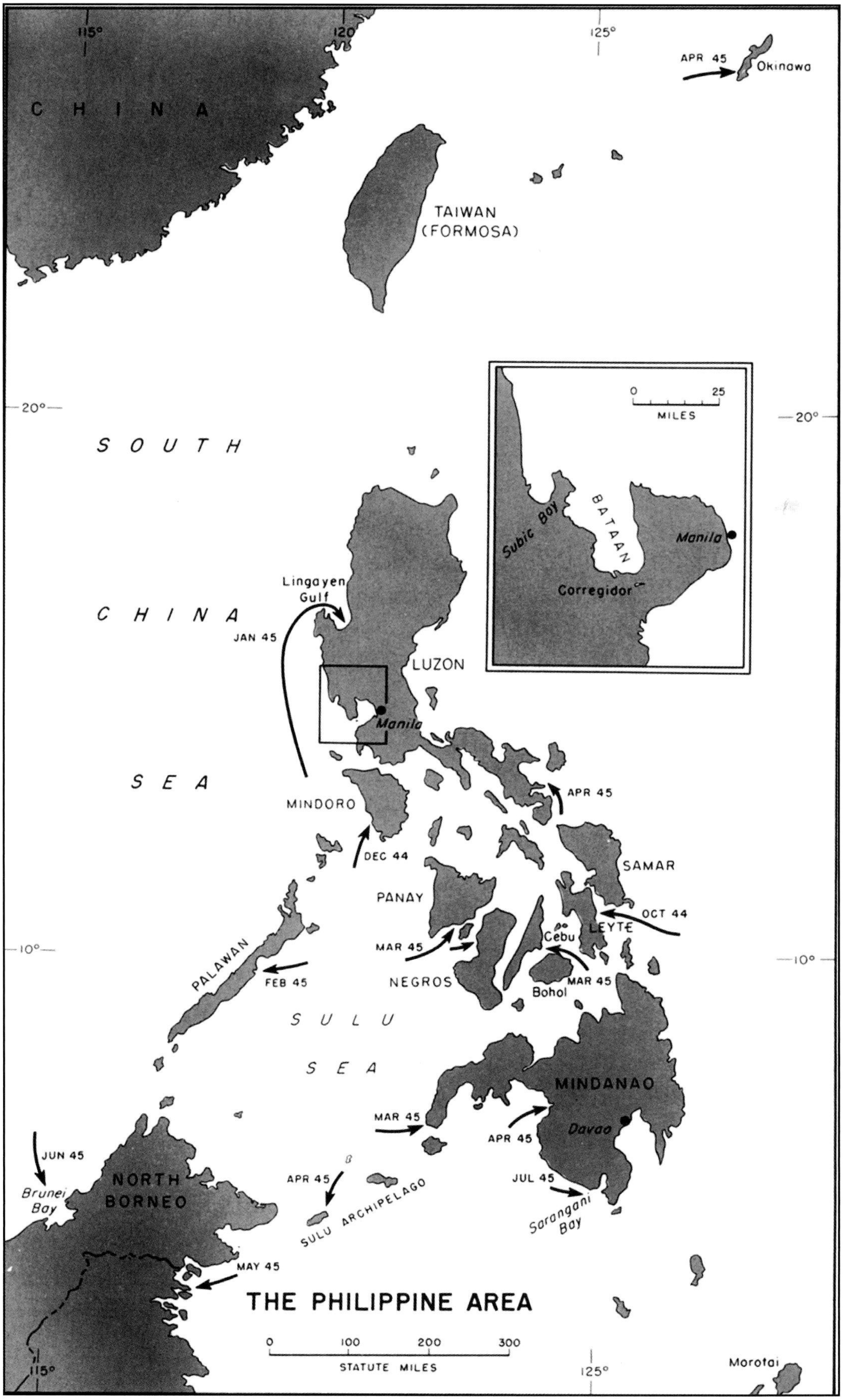

Map of the Philippines. The 13th Air Force night fighters (the 419th NFS and the 550th NFS) defended the central and southwestern areas of the Philippines and covered the Allied landings in Borneo. *Courtesy of the Center for Military History.*

419th NFS Pilot First Lieutenant Clifford Johnson. *Courtesy of AFHRA.*

artillery. Even today, military controllers calculate an imaginary box in which artillery shells will be flying in order to keep friendly aircraft out of the area and away from the danger of being struck by an outgoing friendly round.

The Allied offensives followed two parallel lines stretching southwest from the main body of islands in the Philippines, and the 419th NFS and 550th NFS found themselves split again into separate detachments to support these operations as the United States seized Filipino territory from the Japanese and shifted the campaign to Borneo. The 419th NFS Headquarters Detachment transferred from Middleburg Island to Palawan Island; Detachment A shifted northwestward from Morotai to Zamboanga, located on a peninsula on the southwest side of Mindanao Island; and Detachment B, comprised of aircraft from both the Headquarters Detachment and Detachment A, established itself at Sanga Sanga, just east of Borneo. The 550th NFS Headquarters Detachment flew from Tacloban, on Leyte, while the squadron also dispatched detachments to Zamboanga, Palawan, and Sanga Sanga.

Palawan (March 6 – 27 July 1945)

The 186th Regimental Combat Team of the U.S. Army's 41st Division landed on the Philippine island of Palawan on February 28, 1945, following two days of air and naval bombardment. Palawan is an island province of the Philippines stretching southwestward from Mindoro Island toward Borneo, in the Dutch East Indies. The island is heavily forested with central highlands and has nearly 2,000 miles of coastline. Occupation of Palawan and capture of its airfields would permit American forces to cut one of the two main lines of Japanese communication between Borneo and the Philippine islands, as well as permit air support for the amphibious operations along the many islands lying along the second line of Japanese communications stretching from Zamboanga on Mindanao through Jolo Island and Sanga Sanga to Borneo. The Japanese recognized this strategy and outlined it in documents collected by the U.S. Army after the war:

419th NFS Radar Observer Flight Officer Conrad Stanski. *Courtesy of AFHRA.*

"Meanwhile, the enemy [the United States] had already launched his campaign to clean out Japanese garrisons throughout the central and southern Philippines. Palawan, the first to feel the impact of this mop-up, had by late February lost much of its importance to the Japanese. In possession of the enemy, however, the island would provide an excellent base from which air domination of the South China Sea could be made more complete."[2]

The Japanese Army offered very little opposition to the American forces when they landed near the island's capital of Puerto Princesa on the central southern coast at 8:30 AM. Two American battalions moved inland along the eastern side of Puerto Princesa's harbor while a third battalion drove north from the middle of the bay. The two Japanese airfields were in American hands by 12:00 PM as Japanese forces withdrew into the highlands along the central part of the island and waited. Serious fighting did not occur until the third day of the invasion, when American elements finally encountered the Japanese. Five days of fighting were required to eliminate most of the island's Japanese defenders. Smaller engagements continued until April, when the surviving Japanese soldiers melted into the thick forests and highlands. American soldiers and Filipino guerillas would continue to hunt these survivors until the end of the war. American casualties in the campaign were light and included 12 killed and 56 wounded. Japanese casualties were considerably higher at an estimated 900 dead.[3]

Palawan is possibly best known in World War II history for a brutal Japanese massacre of American prisoners of war less than three months prior to the invasion and the arrival of the 419th NFS. The 2,000 man Japanese garrison on Palawan utilized the forced labor of approximately 350 American prisoners of war captured in 1942. In 1944, the prisoners were working on the airfield at Puerto Princesa while enduring brutal conditions under their Japanese captors. On December 14, 1944, the Japanese garrison wrongly concluded that an American task force bound for Mindoro was actually an invasion force sailing for Palawan. Approximately 150 American prisoners of war were in three covered trenches when Japanese soldiers dumped gasoline on them and ignited the fuel. Those American soldiers who attempted to climb out of the fiery blaze were shot by the Japanese. Only 11 Americans managed to escape the inferno by leaping off a cliff that ran along the back side of one trench. Most of these survivors swam across the bay, where they were rescued by inmates and guards of the Iwahig Penal Colony. Reportedly, many of the guards and other officials of the penal colony were active in the resistance movement against the Japanese and hid the American survivors of the massacre.[4]

The work to improve the two Japanese airfields near Puerto Princesa to support other American operations and interdict Japanese supply lines began immediately after the initial invasion forces moved inland. However, the endeavor was slowed by the marshy soil of the area and poor condition of the bases. The airfields were heavily cratered by American raids prior to the invasion, and the single concrete runway was constructed of poor material. Despite the obstacles engineers diligently tackled the task, and within days established an airfield that would support the requirements of fighter aircraft. Further improvements would continue to make the strip

Puerto Princesa Airfield (Palawan Island) in the Philippines. *Courtesy of Bob Fowble.*

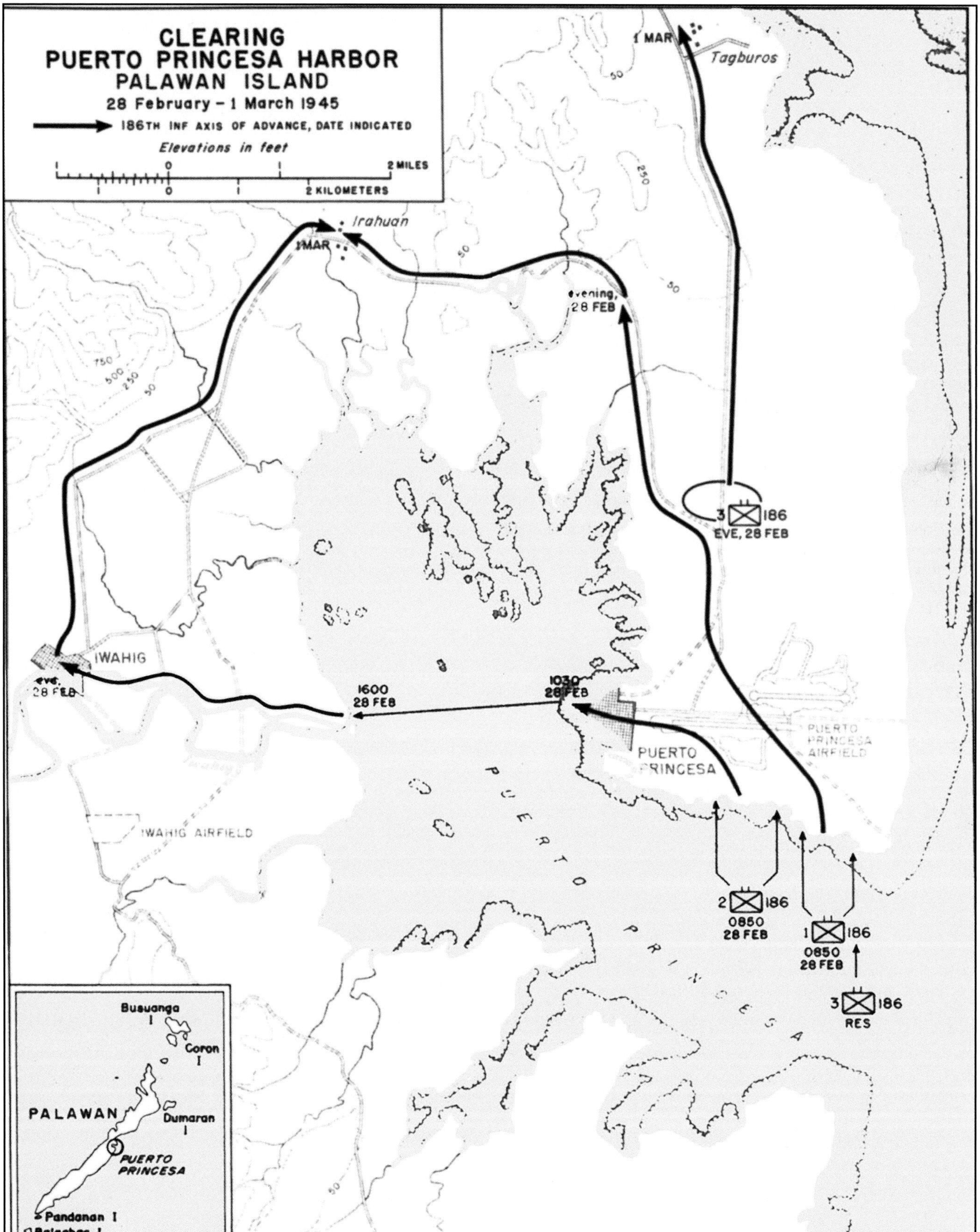

Palawan. *Courtesy of the Center for Military History.*

419th NFS staff at Palawan, in the Philippines. Front row, left to right: First Lieutenant Warren Sloan (S-4), Lieutenant Colonel Norman Jacob (Commander), Captain Weissman (XO), First Lieutenant John Radzieta (Navigation officer), and Captain Warren Slater (S-2). Back row, left to right: First Lieutenant Michael Burlingham (Engineering Officer), First Lieutenant John McGhee (S-3), Captain "Doc" Newman (Flight Surgeon), First Lieutenant "Soup" Campbell (adjutant), Captain James Irwin (Communications and Radar Officer), and First Lieutenant Bob O'Brien (Transportation and Ordnance officer). *Courtesy of AFHRA.*

suitable for bombers. Plans called for the stationing of detachments of the 419th NFS, the 347th Fighter Group (Lockheed P-38 Lightnings), and the 42nd Bomber Group (North American B-25 Mitchells) on Palawan.

Eight days after the initial American invasion, the Headquarters detachment of the 419th NFS (the detachment that had conducted operations on Middleburg Island) transferred from its temporary holding area on Mindoro Island to Palawan in order to support Allied operations in the southwest Philippines, as well as help cut the Japanese line of communications from Borneo Island, located in northern Dutch East Indies. The detachment arrived on Mindoro from Middleburg Island on February 23, 1945, and waited for the invasion of Palawan. On 3 March Squadron personnel departed Mindoro Island on board LST 722 and joined a convoy headed for Palawan, where they arrived on 6 March. The aircraft flew directly from Mindoro to Palawan and joined the ground and spare air crew personnel who had traveled on board the LST. The Black Widows were the first operational aircraft to land at the newly opened airfield. Detachment personnel established the squadron camp two miles north of the air strip in a area of thick brush and coconut tree debris. Several days were required to clear the area and establish a functioning base camp, which was later described by unit personnel as their best since arriving in the theater. The area offered the attractions of top quality beaches as well as wild boars and chickens that supplemented mess rations. Additionally, the area abounded with wild limes, pineapples, and bananas.[5]

The 419th NFS Headquarters Detachment arrived on Palawan with seven P-61s and 41 air crew members, and conducted scramble alerts at night immediately upon arrival on the island. Regular operations commenced on 20 March, three days before the airfield was officially opened for daily air operations. A total of 25 combat missions were flown by the 419th NFS from Puerto Princesa during March, and these included dawn and dusk patrols, night patrols, convoy coverage, sea search and rescue reconnaissance patrols, and scrambles to intercept bogeys detected on radar. Thanks in part to the presence of the 419th NFS, Puerto Princesa did not experience a Japanese air raid during the entire month of March.[6]

The first of the 419th NFS' two March scrambles at Palawan occurred during the early morning hours of 26 March. At 1:20 AM the alert P-61 launched to pursue a bogey reported by the Ground Control Interception (GCI) operator. When the P-61 crew received the intercepting vector and made their turn for an approach, the unknown

419th NFS Enlisted Section NCOs. Front row, left to right: Master Sergeant L. Starkin (Engineering), Technical Sergeant T. Safford (Radar), Sergeant A. Gibson (Operations), First Sergeant R. Lake (Headquarters), Technical Sergeant J. Snaiar (Personnel), Staff Sergeant R. Dunavant (Intelligence), and Master Sergeant G.Wadland (Armament). Middle row, left to right: Master Sergeant J. Bartol (Communications), Staff Sergeant L. Bane (Tech Supply), Technical Sergeant E. Walcher (Armament), Sergeant S. Gillikin (Medics), Technical Sergeant T. Camp (Sheet Metal), and Corporal R. Finizio (Message Center). Back row, left to right: Staff Sergeant N. Zumwalt (Transportation), Staff Sergeant V. Finn (Special Services), Private D. Neigo (Mail Orderly), Technical Sergeant J. Deal (Instruments), Staff Sergeant G. Vandaveer (Supply), Staff Sergeant W. McKinley (Generators and Electrical Maintenance), and Staff Sergeant M. Niedoborski (Mess Sergeant). *Courtesy of AFHRA.*

aircraft also turned to the opposite direction and fled to the south, indicating the possibility the American night fighter radio frequency may have been monitored by an incoming intruder.[7]

Throughout the next evening (26/27 March) the detachment flew six reconnaissance patrols in the search for two unidentified warships reported at the convergence of the Sulu and East China Seas northeast of Palawan. The radar sets on board the P-61s were fairly effective at searching for vessels on the open ocean. The radar wave returns would "paint" the outlines of islands and large vessels in contrast to the wave returns bouncing off the water. The patterns of islands on the radar screens could be utilized in conjunction with a map for nighttime navigation, while smaller returns indicated the possible presence of a large vessel. Reports of the two unidentified warships resulted in six reconnaissance missions of the 419th NFS P-61s as they scoured the ocean with their radar in an attempt to locate the naval bogeys. The Widows did not locate any vessels, and the reports were later annotated as being false. Despite the false reports, the actual reconnaissance patrols were exciting to many personnel in the 419th NFS and served as an acknowledgement that their aircraft could perform a new type of mission against Japanese forces.[8]

The detachment's second March night scramble involved another naval support mission during the evening of 29 March. Reports arrived that a suspicious twin-masted sailing vessel was located east of Puerto Princesa. A 419th NFS P-61 launched from the air strip and successfully located the sailboat. The Black Widow flew cover over and observed the sailboat until a U.S. Navy PT boat arrived on the scene and established that it was a friendly craft.[9]

The atmosphere relaxed around Palawan as the main thrust of the offensive moved toward Borneo, permitting

419th NFS P-61 Black Widow 42-39382 #327. The photo offers a good view of the radar observer's compartment, which provided nearly 180 degrees of visibility behind the airplane. *Courtesy of Don Dessert.*

419th NFS Medical section. *Courtesy of AFHRA.*

419th NFS Orderly Room. *Courtesy of AFHRA.*

squadron personnel to indulge in activities not available since arriving in theater at the end of 1943. Ordnance officer First Lieutenant Bob O'Brien and pilot Dick Stewart opted to go crocodile hunting while on Palawan. The actual hunters were prisoners from the Iwahig Penal Colony, where the survivors of the Palawan Massacre were harbored from the Japanese. The camp tanned and sold the crocodile skins to raise money to support the prison. The two 419th NFS crewmen accompanied several prisoners and a guard to a nearby stream. O'Brien remembered:

"The water was brown with sediment, and the men hunted by diving into the stream and then feeling with their hands for the crocs. The prisoners wore goggles made from the bottoms of glass bottles and would shoot little harpoons under the legs of the crocodiles where the skin was softer. The harpoons were powered by strips of rubber from car inner tubes. The crocs would exhaust themselves by twirling and trying to get free. At that point the prisoners would haul them onto the bank, tie them up, and later take them back to the prison. Stew [Stewart] and I both paid to have a treated skin sent to our homes [in the United States] but neither of us ever got one."[10]

Crocodiles were not the only things hunted by the prisoners. Japanese soldiers were still located in the hills along the central part of the island. Staff Sergeant James Girdis, a radar mechanic, noted some of the prisoners would be temporarily released to hunt for Japanese soldiers. They would return with heads to prove they had found and killed Japanese military personnel still hiding in the hills.[11] Many Japanese soldiers tired of hiding in the hills and running from American military units and vigilantes hunting them. On more than one occasion, Japanese soldiers simply approached American military personnel and offered to surrender despite their original orders to continue their resistance. Sergeant Bob Wright remembered an improbable if humorous scenario when a Japanese soldier actually walked into his tent to surrender:

"I thought he was a Filipino looking for work so I ignored him. Well, he went to the next tent and surrendered. Those guys in the next tent became heroes in the process!"[12]

O'Brien and First Lieutenant George Sloan, the Squadron supply officer, opted out of tent life on Palawan and built their own shack on the beach. The pair acquired some brandy ("horrible stuff from Australia" noted O'Brien years later) and traded it with an Army truck driver for a load of lumber and metal roofing sheet:

"We built a little shack right on the shoreline with a double bunk inside and used sheets for the side facing the ocean, curling up the ends to bring the breeze into the shack. Inner tube rubber was sliced and interlaced to provide a bed covered with blankets for warmth at night. We had it made!"[13]

The locals learned quickly that the newly arrived Americans had money and other items for trade. Staff Sergeant Chuck Pomeroy, a P-61 gunner, noted that individuals who lived in the hills would come down to the American base areas to trade:

"They came to our squadron encampment with blowguns, bows, quivers, and baskets to trade for items of clothing and other things. I later lost a quiver during a hop on a B-25 to Manila. The plane had just taken off from the metal air strip on Palawan when the crew opened the bomb bay doors to rid the aircraft of exhaust and gasoline fumes. I managed to hang on and not fall out but lost the quiver in the process."

The locals would also take any items of value not under guard. Pomeroy remembered that a guard on the airfield fell asleep one night and several parachutes were reported missing the next morning. He noted, "Later I saw Filipino women walking around in nice white nylon dresses."[14]

The Headquarters Detachment P-61s flew 58 combat missions during April 1945, the largest number

419th NFS Medics Tent. *Courtesy of AFHRA.*

Refueling 419th NFS P-61 Black Widow 42-39450 #329. Note the twin stripes on the boom. By the time the 419th NFS moved to the Philippines, the unit moved the three digit numbers on each aircraft from the nose to the vertical stabilizers and added two stripes to each boom. *Courtesy of Leslee Lukas Mays.*

of combat missions for the detachment since September 1944. The 58 missions included 19 to cover minesweeper operations, 12 to provide protection over convoys, and five ocean reconnaissance sorties. The Detachment also performed three of what personnel referred to as "Boy Scout" missions during April. Allied aircraft crews occasionally became disoriented at night and would radio that they were lost. Night fighters would be vectored in the direction of these lost aircraft. The night fighter Radar Observers (R/Os) would then direct their pilots after they acquired the missing aircraft on their radar scopes. After positively identifying the aircraft as Allied and not a Japanese plane trying to sneak into the area as "lost," the night fighter crew would then make contact and direct the plane to follow them to the nearest airfield. In April the Headquarters Detachment located and escorted two lost Navy PB4Y patrol aircraft and one Army P-38.

During April Black Widows from Palawan began staging though the island of Sanga Sanga for operations off the coast of Borneo. The new P-61B models of the Black Widow could be fitted with 165-gallon droppable fuel tanks. The extra fuel permitted the P-61s to depart on dusk patrols from Palawan to Tarakan and then to Sanga Sanga. While on Sanga Sanga the aircraft served as alert planes for night scrambles. At dawn the Black Widows departed Sanga Sanga for their return patrols,

419th NFS Gunner Staff Sergeant Tony Coco.
Courtesy of Don Dessert.

which terminated back at Palawan. The additional fuel in the droppable tanks permitted the P-61s to fly up to 6 ½ hours.

Japanese aircraft made three forays over Puerto Princesa and the American air strip in April. The first, on 21 April, did not result in an attack, although a Japanese aircraft had been reported in the area. The latter two intrusions were confirmed raids resulting in damage to aircraft and injury to American personnel at the air strip. The first of these raids occurred during the early morning hours of 28 April. The Detachment scrambled a P-61 to hunt for two bogeys reported by the CGI operator as approaching Palawan. It should be noted that Japanese pilots were aware of the presence of American night fighters on Palawan and knew of the capabilities and vulnerabilities of the radar systems operated by the aircraft crews, as well as the GCI teams. To counter these systems, Japanese pilots would often approach Palawan and other bases in the area at an altitude that was low enough to permit their arrival with a decreased chance of detection. GCI radar could not detect aircraft flying at very low altitudes due to return clutter. Japanese intruder pilots attempting to avoid the radar by flying at a very low altitude could be spotted if they inadvertently elevated their aircraft. Flying and maintaining a low altitude over the ocean at night without bouncing off the water was a very difficult task, and many pilots would often rise to an altitude that would permit their detection before dropping back below radar coverage. Thus, a radar operator might observe a series of quick radar returns over a period of time, indicating a possible intruder briefly adjusting its altitude as it approached.

Allied forces also maintained picket vessels along the approaches to major bases to spot incoming low altitude Japanese intruder aircraft. It is not clear from the historical records how the American forces knew of the approaching bogeys. Although it was known that two aircraft were attempting to reach the air strip that night, the 419th NFS aircrew and personnel from the Fighter Sector directing their intercept mission were confused about the exact location of the bogeys. The gunner aboard the Black Widow spotted one of the Japanese aircraft and attempted to inform the pilot that he had a confirmed sighting. However, the pilot was talking to Fighter Sector control on the command set at the time and was not able to communicate with the gunner. By the time the gunner got the attention of the P-61 pilot the bogey had disappeared again into the night. One of the enemy aircraft dropped anti-personnel bombs over the north end of the American air strip, wounding two personnel. Both Japanese planes managed to escape the area in the confusion.[15]

The second raid of the month succeeded in damaging several aircraft during the morning hours of 30 April. Two Japanese Mitsubishi G4M Betty bombers (the

419th NFS P-61 Black Widow 42-39473 #330 *Night Vision*. 43-8238 #332 *Vivacious Vivian* sits in the background. Northrop added a communications tube between the pilot's cockpit and the radar observer's compartment in the rear of the Black Widow. In the event of an internal communications failure the pilot could still talk to the radar observer through the tube. *Courtesy of Don Dessert.*

same type of plane shot down by Captain Dick Stewart over Middleburg and First Lieutenant Ralph LeVitt flying from Morotai) approached Puerto Princesa at low altitude along the eastern coast of Palawan. Although the Japanese bombers had been spotted earlier, they currently maintained a low altitude, preventing the GCI operator from vectoring the orbiting Black Widow dispatched to intercept them. At the same time the P-61's R/O could not secure his own radar fix on the bogeys due to their tactics. The Betty bombers managed to reach the air strip, where they commenced strafing and bombing runs over the area, destroying or damaging seven aircraft. This time personnel on the ground did not sustain any injuries. Radio Tokyo bragged about the raid and, according to 419th NFS Assistant Detachment Historian, Second Lieutenant Loren Kane, the Japanese announcer offered "the usual rich imagination and predilection for exaggeration typical of their propaganda."[16] According to Squadron records the Radio Tokyo broadcast announced:

"On the nights of April 28 and 29, large forces of our planes attacked Puerto Princesa air strip on Palawan southeast of Mindoro. Two hundred B-29s, B-25s, and B-24s were destroyed on the ground, and at least 30 P-38s were set afire in addition to numerous military installations hit. The entire air strip was left in flames."[17]

While the numbers of destroyed or damaged aircraft were wildly exaggerated by the radio, it is interesting to note that the Japanese claimed to have destroyed B-29 bombers, which were not based anywhere in the Philippine islands. Squadron personnel speculated that the Japanese bomber crews might have mistaken large PB4Y Flying Boats on the air strip as the giant Boeing B-29 Superfortress bombers.

Morale among the detachment personnel was at its highest level since arriving overseas. Fresh meat, eggs, and butter were served in the mess hall after being flown in from Leyte Island. Wild game occasionally appeared on the mess hall menu. First Lieutenant O'Brien remembered hunting for wild boar with Sergeant Wayne Starnes, the motor sergeant:

Seven gunners of the 419th NFS gather for a group photograph. *Courtesy of Leslee Lukas Mays.*

419th NFS Intelligence Section clerk Sergeant Ralph Dunavant playing ping pong. Squadron members found various ways to entertain themselves when off duty. Films were popular, as well as games, such as softball. In the Philippines, First Lieutenant Bob O'Brien and Sergeant Wayne Starnes opted to borrow carbines and try wild boar hunting. The pair bagged a boar at a wild game trail intersection. O'Brien remembered, "We then carried him back to camp where the cooks made up a meat stew from him that was excellent." *Courtesy of Leslee Lukas Mays.*

"We borrowed carbines and went into the jungle and found an animal trail. The jungle was scary as hell because standing up you can't see 20 feet ahead, but if you stoop way down you can see a great distance and you can pick up trampled paths made by various animals. We got into an intersection of two trails and sat back-to-back with our rifles on full automatic. Soon a boar appeared trotting down the trail and he scared us really well. I fired on him as he came into my gun sight and then Wayne fired on him as well. We both fired, and we could have taken out an enemy squad with the rounds we fired. We killed him. We cut his gut open and emptied it and then bled him. We then carried him back to camp, where the cooks made up a meat stew from him that was excellent."[18]

Mail was arriving from the United States within 10-14 days after being postmarked. For the period this was quite an astounding speed, considering the distances from the United States to the Philippine Islands with so many military personnel deployed in the Pacific Theater during World War II. One has to also consider that aircraft and vessels carrying the mail moved at considerably slower speeds than those of the 21st century. The author served in the Middle East in 2005 and 2006 during Operation Iraqi Freedom. Letters generally arrived there 6-10 days after being postmarked in the United States, so the speed of mail arriving on Palawan during World War II should be considered impressive and a definite boost for morale among military personnel. Morale was also boosted by

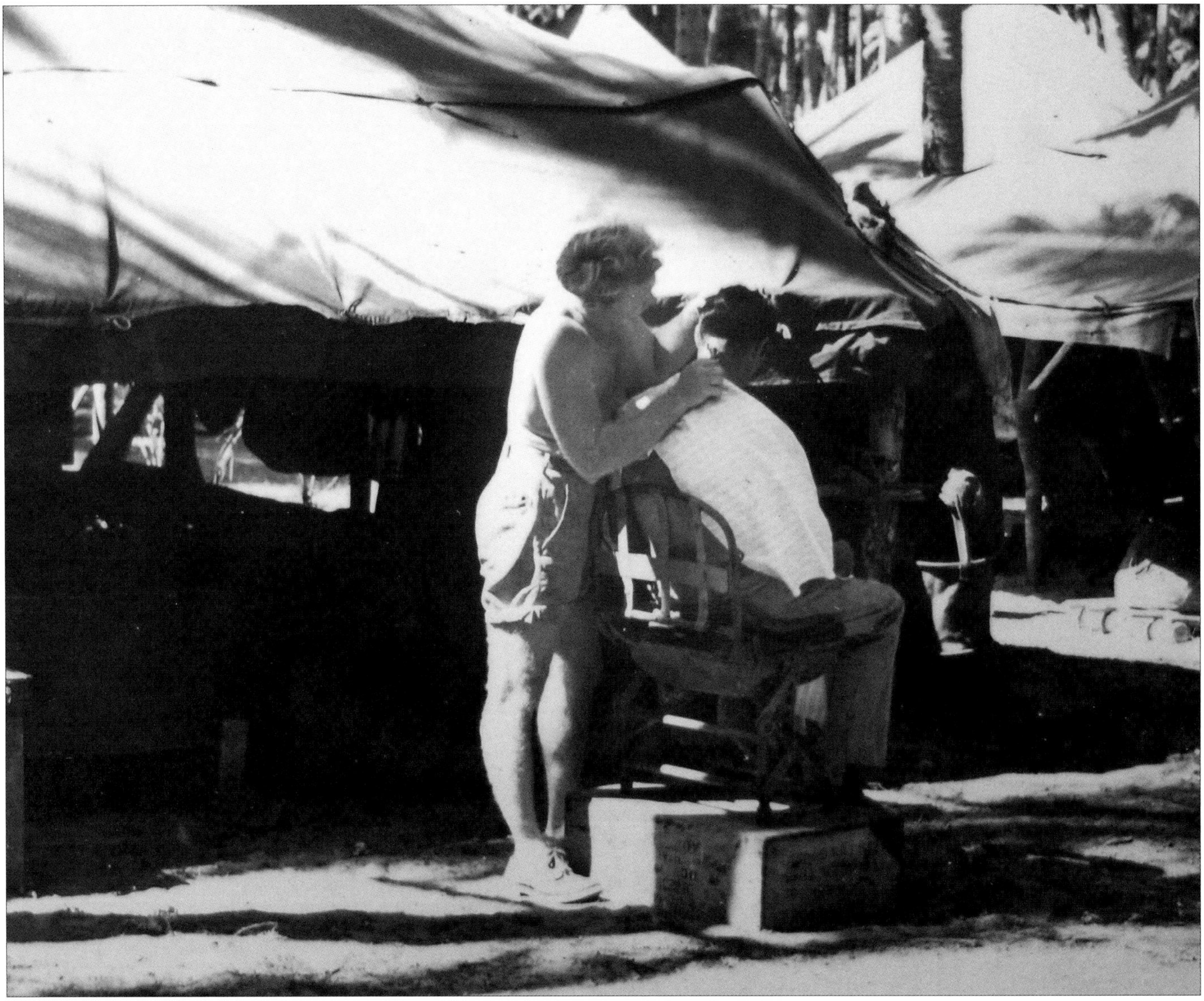

419th NFS barber stand. *Courtesy of Don Dessert.*

the increased beer rations received during this period of the war. 419th NFS historical records brag of each man receiving a 24 bottle supply of beer for the month of April 1945. The small squadron Post Exchange carried a "very adequate" supply of personal hygiene items, candy, fruit juices, magazines, and books. In addition, the military successes of the European Theater of Operations and the impending collapse of Germany meant personnel in the Pacific Theater could expect the transfer of entire units from Europe in the coming months. Although these reinforcements from Europe would not be required due to the surrender of Japan, the thought of significant reinforcements boosted the morale of the military personnel in the Pacific. Simply stated, life was better for the 419th NFS that it had ever been since departing Florida in 1943.[19]

After 15 May the 419th NFS discontinued its night missions to Sanga Sanga. In total, the detachment pilots flew 101 missions during May, including eight night scrambles on Palawan and four night scrambles from Sanga Sanga. Other missions included 38 dawn and dusk patrols, 15 night patrols of Palawan, 38 dawn and dusk patrols over the beach head established at Tarakan, three reconnaissance, and three night convoy escort missions. Japanese raiders approached close enough to Palawan to result in red alerts twice during the month. On 1 May an unidentified bogey was plotted at 110 miles southwest of Puerto Princesa at 3:37 AM, resulting in the scramble of two P-61 night fighters. One Black Widow remained over the air strip while the second flew 50 miles toward the direction of the incoming bogey and waited for guidance from the GCI radar operator. At 4:02 AM radar reported the bogey now at 45 miles northeast of Puerto Princesa. Although a red alert was sounded to warn personnel on the air strip the bogey never approached the airfield. One P-61 landed and the other remained airborne as a precaution.[20]

The second red alert in May involved an intercept attempt and brief engagement between a P-61 and Japanese Aichi D3A Val. The Val fighter, the same type of aircraft pursued by First Lieutenant Guy Scheetz near Middleburg on November 9, 1944, was originally a carrier-based dive bomber that had been converted to other duties, including night intruder, late in the war. During the early morning hours of 27 May two bogeys approached Puerto Princesa from the southwest. Japanese aircraft frequently flew toward Palawan from the southwest because they were launched from bases on Borneo. One P-61 was already on patrol in the area, and it was vectored toward the two bogeys while a second P-61 was scrambled to cover the airfield. The first P-61 received word that the GCI radar had pinpointed the bogeys at 45 miles away. The GCI operator then provided a vector for the Black Widow crew. The P-61 R/O picked up the bogeys on his radar scope and began directing the pilot, Captain James Lane, toward the intercept. The bogeys, identified as Vals, knew they were being stalked and flew erratically to force the P-61 to lose radar contact. The R/O continuously lost and regained radar contact as the Vals flew closer to Puerto Princesa. It should be noted here as a reminder that the R/Os not only had to contend with identifying the direction of the bogeys but also the altitude. Savvy Japanese pilots knew to continuously change their heading and altitude to confuse pursuing American night fighter crews.

As one Val maneuvered closer to Puerto Princesa, the Fighter Sector control personnel ordered the P-61 to discontinue pursuit so that American anti-aircraft artillery could engage the Japanese aircraft and not put the Black Widow in danger. The Val successfully flew through the anti-aircraft fire and crossed the airfield at 4,000 feet, dropping anti-personnel bombs. Seven personnel were wounded, but all aircraft on the ground escaped damage in the attack. The Val then flew toward the southeast,

419th NFS motor pool and dispatcher's shack. *Courtesy of Don Dessert.*

Japanese Aichi D3A Val dive bomber. P-61 Black Widow pilot Captain James Lane of the 419th NFS dueled with a Val during May 1945. Lane's Black Widow was one of the late "A" models, which did not have an upper machine gun turret but carried only the 20mm cannon in fixed mounts underneath the aircraft. The lack of an upper gun turret meant that the pilot was responsible for not only flying, but also operating all of the aircraft armament and aiming the guns by turning the plane. The Val pilot managed to drop anti-personnel bombs across the American airfield, but was intercepted by Lane as he tried to depart. Lane and his radar observer located, fired upon, and then lost the Val three times before the Japanese plane completely disappeared. After the engagement, Captain Lane commented that he was certain that the crew would have downed the Val if their plane carried an upper gun turret and a gunner. The early "B" P-61 models were also turret-less before Northrop added the machine gun position in later "B" models. *Courtesy of Don Dessert.*

where the P-61 awaited. GCI radar provided a vector and the P-61 maneuvered for an intercept. The P-61 R/O made his own radar contact and directed Lane for a turn to attack the Val. The Black Widow crew spotted the exhaust flames from the Val's engine and moved closer for a kill. The Val pilot knew he was being stalked and took evasive action.

The 419th NFS P-61 flown by Captain Lane in the engagement with the Japanese Val was one of the late "A" models, which did not have an upper machine gun turret but carried all of its weapons in fixed mounts underneath the aircraft. The lack of an upper gun turret meant that the pilot was responsible for not only flying, but also operating all of the aircraft armament. The early P-61As issued to the 419th NFS in mid-1944 included the gun turret and a position for a gunner. On these aircraft, the gunner operated the four .50 caliber machine guns in the upper turret while the pilot controlled the 20mm cannon mounted under the belly of the aircraft. A turret allowed the crew to maneuver the guns and fire at an aircraft at any compass point despite the direction being flown by the pilot. In a twisting dog fight, control of the machine gun turret could be electronically passed to the R/O in the rear, who could then fire upon enemy aircraft behind the P-61. Without the upper turret and a gunner, the P-61 pilot had to fly the aircraft and operate the weapons systems. The guns operated by the pilot were in a fixed position like other fighters. Thus, the pilot had to aim his weapons by turning his plane toward the direction in which he wanted to fire. Unless the attacking aircraft was directly behind an enemy plane, a pilot or gunner had to consider deflection when firing a weapon. With a moving target one does not fire at the enemy aircraft, but where it should be when the bullets reach the moving plane. The speed of an enemy aircraft along with its distance and direction in a turn determined the deflection, or angle, at which a bullet must be fired to strike the aircraft. While this can be difficult for any pilot, it's even more challenging for a night fighter pilot, who has extremely limited vision of his target.

Captain Lane turned his aircraft and fired a long 20mm cannon burst with a 90-degree deflection. The Val broke contact, dove for the ocean to lose radar contact, and disappeared. Ten minutes later the P-61 crew again made visual identification of the Val at 3,000 feet and closed for a second attack. The pilot fired a two-second 20mm cannon burst with a 30 degree deflection. Again the Val took evasive action, dived toward the ocean, and disappeared. Five minutes later the P-61 crew made a third visual contact and closed for another attack. The P-61 pilot fired another two-second burst, forcing the Val to again dive toward the ocean and fly into low lying clouds 45 miles southeast of Puerto Princesa. The crew was not able to determine if the Val sustained any hits and the Japanese aircraft was not seen again that night. After the engagement, Captain Lane commented that he was certain that the crew would have downed the Val if their plane had an upper gun turret and carried a gunner. The second Val bombed Books Point, on Palawan, and managed to escape into the night. A mission critique session the next morning between 419th NFS air crews and GCI controllers of the 320th Fighter Control Squadron resulted in a determination that the intercept was handled successfully and that the interception would have resulted in a kill if the P-61 had carried a turret and gunner.[21]

Although May beer rations were only half of those received in April and the number of days with fresh meat, eggs, and butter were also decreased due to the movement of the supplying Quartermaster unit on Leyte, morale was still fairly high for detachment personnel during the month. Three day passes to Manila were the highlight of the month with transportation provided

419th NFS Intelligence Section. Captain Warren Slater sits in the foreground while Sergeant Ralph Dunavant can be seen at his typewriter in the background. *Courtesy of Don Dessert.*

by the unit Douglas C-47 Skytrain. However, due to maintenance issues with the C-47 only one flight of eleven personnel made it to Manila during the month. Germany fell in May, leading to many conversations about when everyone could finally go home.

The Headquarters Detachment on Palawan flew 92 combat missions in June 1945 and scored another confirmed kill for the 419th NFS. The missions included 35 dusk and dawn patrols over Palawan, four night patrols across Palawan, 15 patrols over Brunei on the island of Borneo, two night intruder missions, 34 convoy coverage patrols, and two minesweeper coverage patrols. New P-61 aircraft continued to arrive in the theater, and during June the Headquarters Detachment exchanged two war weary P-61A model aircraft for new P-61Bs, and also received an additional two P-38s. While Lieutenant Colonel Norman Jacob remained the 419th NFS commander, Captain Ralph LeVitt served as the Headquarters Detachment commander.

Australian forces invaded Tarakan Island, Borneo (located at the southwest end of the Sulu Archipelago stretching from Zamboanga to Borneo), on May 1, 1945. Allied air support for the operation originated from Morotai as well as Palawan. The 419th NFS planes flying cover over Tarakan were the aircraft staging through Sanga Sanga from late April until 15 May. On 10 June Australian troops landed at Brunei Bay, on the northwest tip of Borneo, covered by Allied air and sea power. Three days later, the Headquarters Detachment of the 419th NFS based on Palawan provided the night fighter protection for the invasion.

419th NFS Pilot First Lieutenant Harold Michels (right) receives an Air Medal from Squadron Commander Lieutenant Colonel Norman Jacob. Michels and his crew bagged a Japanese Kawasaki Ki-45 Nick fighter over Borneo in June 1945. *Courtesy of AFHRA.*

Coverage of the Allied operation at Brunei Bay resulted in the last confirmed kill of the 419th NFS. First Lieutenant Harold Michels bagged the 419th NFS' last confirmed kill of the war on the evening of 13 June. Michels launched around 4:00 PM from Puerto Princesa airfield bound for a dusk patrol over Brunei Bay, which lay approximately 2 hours flying time southwest of Palawan. The pilot flew a newly arrived P-61B model which had a total of only 30 hours flying time. For the long distance mission four 165-gallon droppable auxiliary fuel tanks were added to the aircraft to extend her range and time aloft on the patrol. Michels and his crew, Flight Officer Verna J. Morgan (R/O) and Sergeant John C. Snyder (gunner), arrived over Brunei Bay around 6:00 PM and initiated their patrol. An hour later a shipboard GCI operator working with the aircrew reported a contact on his radar system. The GCI operator instructed Michels to orbit his aircraft at 6,000 feet and await further instructions.

Four minutes later the GCI operator ordered the P-61 crew to assume a vector of 080 degrees. Another minute later Michels recalled receiving another radio transmission from the GCI operator informing him:

"a bogey 10 miles approaching from the south. About 30 seconds later I made a visual of the bogey at one mile, 30 degrees to starboard, 1,000 feet below at 5,000 feet. My gunner confirmed the visual. At the same time my radar observer announced a radar contact."

Michels turned the P-61 to starboard and began his pursuit. The bogey spotted the American Black Widow and banked west into the setting sun in a shallow dive to shake the pursuit. Michels recorded later that the sun "was a big, full-circled orange glow" and the Japanese pilot wanted to disappear into its glare.

The P-61 crew continued their pursuit and pulled to within one-half mile behind the bogey. Michels noted:

Japanese Kawasaki Ki-45 Nick fighter. The Japanese utilized the Ki-45 Toryu (Dragon Slayer), known by the Allied codename Nick, in many combat roles, including that of a defensive interceptor against American heavy bombers. The Japanese employed Nicks operating near the Philippines in 1945 as intruders against Allied naval vessels. The twin-engine fighter carried a crew of two men. In June 1945, First Lieutenant Harold Michels and his crew, Flight Officer Verna J. Morgan (radar observer) and Sergeant John C. Snyder (gunner), bagged a Nick that was attempting to intercept American naval vessels near Borneo. Michels and Snyder traded shots with the Nick's rear gunner as defensive fire streaked past the P-61 Black Widow. The Nick's starboard engine caught fire as sections of both wings flew off the plane. The P-61 crew continued firing at the Japanese aircraft and watched as the Nick flew straight into the ground and exploded. *Courtesy of Don Dessert.*

"I closed rapidly at full power in a shallow dive and 15 to 20 seconds later was dead astern of the then recognizable Nick at 500 yards."

The Nick, the Allied codename for the Japanese Kawasaki Ki-45 Toryu (Dragon Slayer), originated as a long range bomber escort fighter early in the war. The Nick was a two-seat, twin-engine aircraft that evolved into an interceptor for American B-29 bombers over Japan. In the Southwest Pacific the Japanese often employed the Nick for anti-ship operations. The aircraft could carry one 37mm and two 20mm cannons, as well as wing racks for two 550 pound bombs. During the Allied Brunei Bay operation, Japanese Nicks were out at night seeking targets of opportunity among the Allied vessels in support of the invasion. The mission of the 419th NFS fighters covering Brunei Bay included intercepting any aircraft attempting to attack the fleet.

The Nick bore a standard Japanese mottled grey, green, and brown camouflage according to the P-61 crew. Michels opened fire with the 20mm cannon as Snyder did the same with the turret mounted .50 caliber machine guns. The rear gunner in the Nick returned fire, and his tracers flashed past the Black Widow crew without scoring any hits. However, the rounds from the P-61 were more accurate, and Michels remembered seeing strikes on the starboard side of the Nick. Smoke poured from the Nick's starboard engine. Michels turned his aircraft with the Nick and lined up a second shot on the Japanese plane. He moved within 350 yards, and the crew opened fire again with the 20mm cannon and .50 caliber machine guns. Rounds pelted the already wounded starboard side of the aircraft. The Nick's rear gunner again returned fire with no results. Suddenly, the Nick's starboard engine caught fire and burned fiercely as the plane continued its dive. Sections of both wings flew off the Nick as the P-61 continued firing at the Japanese aircraft. The Nick dived straight into the ground and exploded about 300 yards inland from the coast approximately 15 miles south of Brooketon. Michels and his crew received a congratulatory radio call from two American PT patrol boats and a destroyer that witnessed the kill and confirmed the downing.

After the engagement Snyder reported that he could not turn his turret. During development of the P-61 Northrop encountered numerous problems with the upper machine gun turrets, including jamming, so they would not turn in flight. Another problem involved excessive buffeting as the air stream passed over the aircraft. The latter caused the aircraft to shake and could prove dangerous in flight. Many 419th NFS crews experienced jammed gun turrets during combat operations. As a result, Northrop discontinued adding gun turrets to the latter aircraft produced under the P-61A series until they could solve the problems. In addition, the P-61 carried the same type of gun turret as the B-29 bomber. The military designated the latter aircraft as a priority recipient for the turrets, and this became another factor in the decision to remove the gun turrets in the later P-61A models. The P-61 flown by Captain Lane when he pursued the Japanese Val was one of these later A models without a turret. After production began on the P-61B series aircraft the military gave approval to divert improved turrets to the night fighter. Northrop believed that it had corrected the various malfunctions experienced with the gun turret on the A series aircraft. Despite the improvements made with the P-61B model and its new turret, the gun platform still tended to fail in combat. Michels reported turret trouble to the GCI operator and was relieved of his patrol by a second P-61 in the area flown by First Lieutenant Robert Frenzel, along with his R/O First Lieutenant George Black, who had been on an intruder mission over northeast Borneo. Frenzel and his crew had not encountered any Japanese aircraft during the operation and agreed to complete Michels' mission. Michels and his crew returned to Palawan and landed around 11 PM to a warm welcome from the ground personnel.[22]

Four days later First Lieutenant Frenzel flew the longest operational mission of a P-61 during World War II—a total of 9 hours and ten minutes. Frenzel, along with R/O First Lieutenant Francis Meyer and gunner Private First Class Thomas Doyle, launched their 1,600 mile night intruder mission on the afternoon of 17 June. The 419th NFS had shifted from night intruder missions after receiving their initial P-61s in the summer of 1944. Frenzel's mission would demonstrate the versatility of the P-61 Black Widow for this type of combat operation. The crew flew to Kushing, in southwest Borneo, in search of Japanese aircraft. Frenzel spent nearly two hours over the target area—two Japanese airfields. Poor visibility in the area thwarted the efforts of the P-61 crew to stalk and kill an enemy aircraft attempting to utilize one of the airfields. Frenzel returned to Palawan, where he also encountered poor weather and limited visibility, necessitating a successful instrument landing at the American air strip. Although the crew did not bag any Japanese aircraft during the mission, they did prove that long intruder operations were possible with the Black Widow.[23]

The war continued to push southwest of Palawan as operations increased on Borneo and its associated small islands. Black Widows of the 550th NFS arrived on 9

Left to right: First Lieutenant John McGhee (pilot) and First Lieutenant Francis Meyer (radar observer). Meyer flew as the radar observer for First Lieutenant Robert Frenzel (pilot) and Private First Class Thomas Doyle (gunner) in June 1945 when the trio completed the longest operational mission of a P-61 during World War II—a total of 9 hours and ten minutes. The mission demonstrated the versatility of the P-61 Black Widow for this type of long range combat operation. The crew flew to Kushing, in southwest Borneo, in search of Japanese aircraft. *Courtesy of Don Dessert.*

419th NFS Radar Observer Flight Officer Richard Bell. *Courtesy of AFHRA.*

June and flew operations from Palawan until 19 June, when they transferred to Sanga Sanga. In response to the need for closer night fighter support, the 419th NFS formed a Detachment B as requested by 13th Air Force for night fighter coverage from Sanga Sanga and joined the 550th NFS aircraft there. Detachment B comprised men and aircraft from both the Headquarters Detachment on Palawan and Detachment A at Zamboanga. Men and equipment began the transfer from Palawan on 20 June. Squadron Commander, Lieutenant Colonel Norman Jacob, left with the Palawan group headed for Sanga Sanga. Three day passes to Manila continued in earnest until they were cancelled due to sanitary concerns within the city. Despite the loss of the pass opportunity, life on Palawan continued to be better than the unit had experienced at other locations. Increased amounts of fresh food from Leyte and a long awaited ice machine arrived to the joy of all personnel. With the latter came a long anticipated delight—ice cream, which was served almost daily during the last week of June.

Following the transfer of nine P-61s to Sanga Sanga in June, the 419th NFS Headquarters Detachment continued to operate from Palawan with two P-38s, as well as the Squadron's C-47 until 22 July, when their P-61s returned from Sanga Sanga. Four of the new P-61B aircraft arrived during the month to replace war weary P-61As. July was a quiet month for the Headquarters Detachment, since the war had tended to continue pushing southwest of the island. There were no red alerts, air raids, or night fighter scrambles during the month. Before the end of the month Detachment A moved to Palawan from Zamboanga, and all squadron elements on Sanga Sanga also joined the Headquarters Detachment at Puerto Princesa. This was the first time since June 1944 that the entire 419th NFS

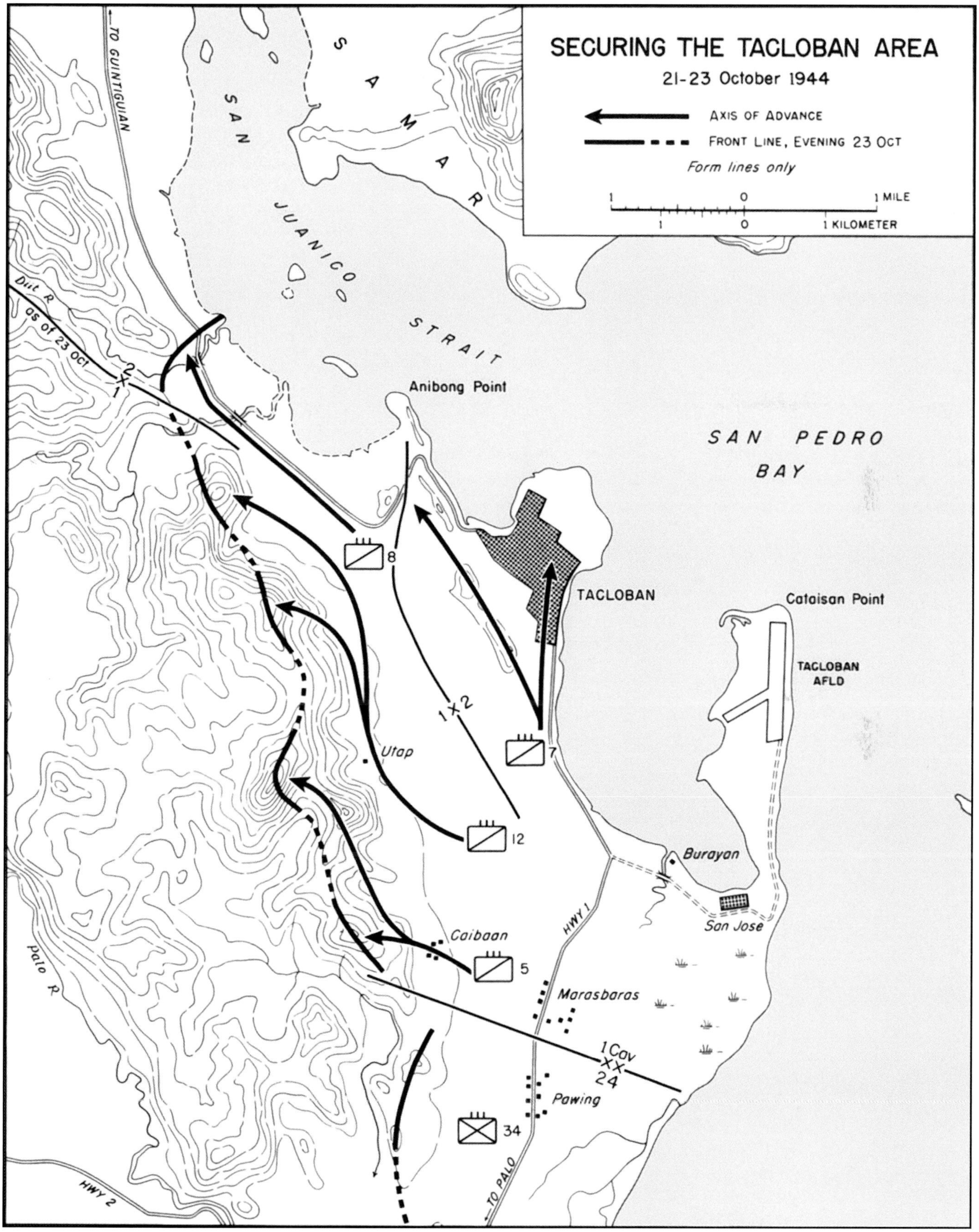

Tacloban and Tacloban Airfield. *Courtesy of the Center for Military History.*

could be found at one location.[24] With the unit reunited and the operations to the southwest of the Philippines winding down, preparations for the invasion of Japan were now the focus of the 419th NFS.

Tacloban (March 8 1945 - 31 July 1945)

The U.S. Sixth Army, commanded by Lieutenant General Walter Krueger, consisted of two Corps sized elements, with each fielding two divisions. Major General Franklin C. Sibert's X Corps, consisting of the 1st Cavalry Division and the 24th Infantry Division, was designated to land on the northeast corner of Leyte at Tacloban. The Japanese 16th Division under Lieutenant General Shiro Makino opposed the Sixth Army's landings on Leyte. The X Corps landed on October 20, 1944, and moved inland, while the XXIV Corps came ashore on Leyte south of the X Corps area. The 1st Cavalry Division seized the town of Tacloban and its airfield.

Five 550th NFS Black Widows and eight C-47 Skytrains carrying the ground personnel flew into the airfield at Talcloban on March 8, 1945. The squadron immediately assumed its night fighter duties and began patrolling the next day. The 550th NFS P-61s were on hand to provide coverage for the American forces landing at Zamboanga on 10 March. Upon the arrival of a detachment of the 419th NFS at Zamboanga on 16 March the 550th NFS turned to supporting the military operations at Iloilo, on Panay. At the same time two Black Widows provided convoy coverage in the Mindanao sea area. Severe weather conditions in mid-March were responsible for forcing four different P-61s to make emergency landings at Dipolog, on Mindanao, which was held by Filipino guerrillas. Two landed on 12 March and two more Black Widows diverted there the next day, leaving only one 550th NFS P-61 to provide night fighter coverage out of Talcloban. On 14 March twelve C-47s arrived at Dipolog and brought fuel for the stranded Black Widows to return to their web at Talcloban.

On 19 March Major Robert Tyler, the commander of the 550th NFS, was on patrol when he received a

Damage to the wing of a P-61 Black Widow flown by First Lieutenant Donald Coski of the 550th NFS. Coski reported being hit during an intruder mission and landed at a recently captured auxiliary airfield. Vegetation reportedly caused the damage, but it is not clear whether the damage resulted from the low level attack or the landing. The 550th NFS spent most of its time in the Philippines flying intruder missions. Being relegated to intruder missions in an area of decreasing combat bothered the 13th Air Force night fighters, who since mid-1944 watched their counterparts in the 5th Air Force move forward into areas contested by greater numbers of Japanese planes, resulting in more American kills of enemy aircraft. *Courtesy of AFHRA.*

550th NFS P-61 Black Widow piloted by Second Lieutenant Melvin Stevick. During April 1945, 550th NFS commander Major Robert Tyler and Stevik flew their P-61s on a routine harassment mission against the Japanese. The pair were directed to strafe an area in which a Japanese convoy recently passed. Stevick reported, "...On our second pass we hit something on the east side of the town that exploded lighting up the entire area. Debris was seen mushrooming to about 1,200 feet. We strafed one large truck about two miles east of town on the highway leading to the hills...If we had been ordered over the area fifteen minutes earlier, we would have completely destroyed a large convoy of men and equipment retreating into the hills." *Courtesy of AFHRA.*

call from a Filipino guerrilla GCI station calling for assistance. The GCI operator identified a Japanese target and requested Tyler to strafe it. Tyler complied and attacked a Japanese bivouac area on Cebu Island. The pilot made six passes over the area, expending over 500 rounds of ammunition before landing at Iloilo. A second P-61 strafed another target identified by the GCI operator. The next morning, Tyler ate breakfast with a Filipino guerrilla leader named "Colonel Alemdras" who briefed him on the Japanese situation in the area. Upon departing the airfield Tyler strafed another target identified by the guerrillas.[25] Missions in support of the Filipino guerrillas continued for the rest of the squadron's assignment at Talcloban.

On 20 March Second Lieutenant Raymond Ont (pilot) and Second Lieutenant John Chervenell (R/O) flew the 550th NFS' first night intruder mission. Major Tyler scrambled in his P-61 on 25 March for what had become a rare intercept attempt against a bogey. The R/O made contact with the bogey over Ormoc Bay at 9:00 PM, and Tyler banked the Black Widow to starboard to initiate a pursuit. The bogey dove to port and Tyler followed. "We followed and hit his prop wash very hard. We then lost contact and never picked him up again," noted Tyler in his report.[26] "Prop wash" is a pocket of unstable air caused by the propellers of an aircraft. A pilot flying very closely behind the first aircraft and entering this pocket can experience temporary buffeting, leading to diminished control of the plane.

The only confirmed destruction of a Japanese aircraft by the 550th NFS, although it was on the ground, occurred on 30 March. Second Lieutenant Edward Jiru spotted a camouflaged Japanese Nakajima Ki-43 Oscar at Tasbilaran air strip. Jiru wrote in his mission report:

"The [Japanese] aircraft was located on the edge of a small clump of trees. The camouflage seemed to be in the nature of a grass covering over the aircraft. I made one pass and the aircraft blew up and burned."

Major Victor Modena (in the round hat) stands with other members of the 550th NFS beside his P-61 Black Widow 42-39485 *The Gay Blade*. *Courtesy of AFHRA.*

Jiru strafed buildings at the air strip after destroying the Oscar and expended 640 rounds of 20 mm ammunition in the attack.[27] At the conclusion of March, the 550th NFS had flown 138 missions for the month.

Most of the 550th NFS missions in April were day and night harassing operations. Pilots Major Tyler and Second Lieutenant Melvin Stevick flew two P-61s on a typical harassing mission on 2 April. Tyler and Stevick took off from Talcloban and arrived in the patrol area by 6:00 PM. GCI contacted the pilot and directed him to strafe an area around Cuimbalaon. Stevick reported:

"Large fires were started on the east side [of Cuimbalaon], and on our second pass we hit something on the east side of the town that exploded, lighting up the entire area. Debris was seen mushrooming to about 1,200 feet. We strafed one large truck about two miles east of town on the highway leading to the hills. Hits were observed on ground installations starting fires. If we had been ordered over the area fifteen minutes earlier we would have completely destroyed a large convoy of men and equipment retreating into the hills."

The pilot expended 450 rounds of 20 mm ammunition in the strafing attack. Tyler concurred with Stevick's comments and declared:

"We saw about seven vehicles and managed to hit two of these and a nearby house. If we had been given these targets fifteen minutes earlier, we could have wiped out the whole bunch."

Tyler utilized 400 rounds of 20 mm ammunition.[28]

The next day Major Victor Modena flew a similar mission as the one reported by Tyler and Stevick. Modena was on an early morning patrol when directed to a target area:

"We strafed, destroyed, blew up, and otherwise damaged 1 ack-ack [anti-aircraft] truck, 1 staff car, 1 full oil truck, 1 cargo truck, 2 light trucks, 2 recon [reconnaissance] cars, and one general dump. Upon the arrival of B-24s and Corsairs [Chance Vought F4U] on the scene I directed them to a motor pool in the Japanese area," wrote Modena after the attack.[29]

On 5 April the 550th NFS nearly lost an aircraft piloted by Second Lieutenant Stevick. The pilot was strafing three anti-aircraft gun positions when he sustained a direct hit from a 40 mm shell in the port wing. He managed to stabilize the aircraft and come down at the emergency air strip at Iloilo, where he blew a tire during the landing.[30] The squadron dispatched four Black Widows and eight air crews on 27 April to Zamboanga, where the 419th NFS maintained a detachment of P-61s.

May was a relatively quiet month for the 550th NFS. The first half of May was devoted to providing coverage for the American landings in northern Mindanao. The squadron made only one scramble during the month to investigate a bogey. On 11 May the detachment at Zamboanga returned to Tacloban. Morale among the squadron personnel was reported to be fairly high at this period. The most important morale booster reported in the monthly squadron history was the addition of coca cola dispensers in the enlisted and officers' clubs. However, the lack of ice for the drinks was bemoaned, as the squadron personnel sought ways to acquire more supplies.[31]

June proved to be uneventful for the 550th NFS as well. The squadron flew three scramble missions from Tacloban in search of reported bogeys during the month. Two of these missions were flown in P-38 Lightnings maintained by the squadron for daylight defense. Six P-61s transferred to Zamboanga on 1 June and four Black Widows shifted to Palawan on 9 June. The 419th NFS maintained a detachment at both locations. On 20 June the 550th NFS aircraft and crews at Palawan moved again to Sanga Sanga to join the other 550th NFS Black Widows from Zamboanga. The most significant activities at Tacloban during June included the testing of air-to-ground rockets attached beneath the wings of the P-61s.

Four rockets mounted under the wing of a 550th NFS P-61 Black Widow. Note the external fuel tank between the rockets and aircraft fuselage. Long range patrols from bases in the Philippines demanded supplemental fuel. *Courtesy of AFHRA.*

Installing rockets on a 550th NFS P-61 Black Widow. The 550th NFS experimented with rockets for use in the intruder role against Japanese targets in the Philippines. The 419th NFS also experimented with rockets. *Courtesy of AFHRA.*

Although the 550th NFS lacked a sufficient number of mechanics for its aircraft, the June unit history notes, "The most serious shortage that we have had to contend with is the lack of refrigeration in the mess hall."[32]

During July, the 550th NFS provided air defense support for the area in cooperation with the 339th Fighter Squadron and later the 12th Fighter Squadron. Although not well recorded, the squadron lost a P-38 to a landing accident at Tacloban in July. The pilot survived the accident. The squadron detachment at Sanga Sanga departed on 31 July and returned to Tacloban. Morale at Tacloban remained high for the month of July. Dishes and trays arrived for the squadron mess halls that:

"greatly increased the already existing high morale… The problem [from June] of no refrigeration was further felt during the period when fresh vegetables and fruit were received from our first 'Fat Cat' trip to Australia. This supply of fresh vegetables was another important morale factor, for it was the beginning of regularly scheduled trips to Australia for foodstuffs which are not to be obtained in our regular Quartermaster rations."[33]

Zamboanga (March 16 – 20 July 1945)

Approximately 9,000 Japanese soldiers of the 54th Independent Mixed Brigade garrisoned the Zamboanga Peninsula, which stretches from the west end of the Philippine island of Mindanao. Zamboanga is located on the opposite (southern) side of the Sulu Sea from Palawan. Seizure of Zamboanga and its airfield was considered critical for the establishment of air operations to support the southwest march of American forces along the Sulu Archipelago through Jolo Island and Sanga Sanga to cut the Japanese links between the Philippines and Borneo. The Japanese also recognized

550th NFS pilot in one of the squadron's P-38 Lightnings. *Courtesy of AFHRA.*

the importance of Zamboanga to the American offensive and noted in documents compiled after the war:

"On the heels of the invasion of Palawan, other American forces swung south to invade Zamboanga and the Sulu Archipelago. Capture of the air and naval bases along this strategic link between the Philippines and Borneo would complete the isolation of the Philippines from the Japanese-held areas to the south."[34]

The Japanese forces were concentrated around the town of Zamboanga and the southern tip of the peninsula. Air support from the new American airfield on Palawan was envisioned as an important element in the operation to seize Zamboanga. The delays in its construction jeopardized the offensive until Filipino guerrilla forces seized a small Japanese air strip 145 miles northeast of Zamboanga City. Marine Corps aircraft flew into the strip and utilized it to cover the landings in cooperation with attacks by the 13th Air Force from other airfields. The 162nd and 163rd Regiments of the 41st Infantry Division landed near Zamboanga City on 10 March. Japanese troops offered little resistance at the beachhead and withdrew into the hills in a tactic similar to that of the defenders on Palawan. Two weeks were required to break Japanese resistance in the hills running along the spine of the peninsula and force them to begin a withdrawal northward, where they were continually harassed by Filipino guerillas. Operations by American and Filipino forces continued for several

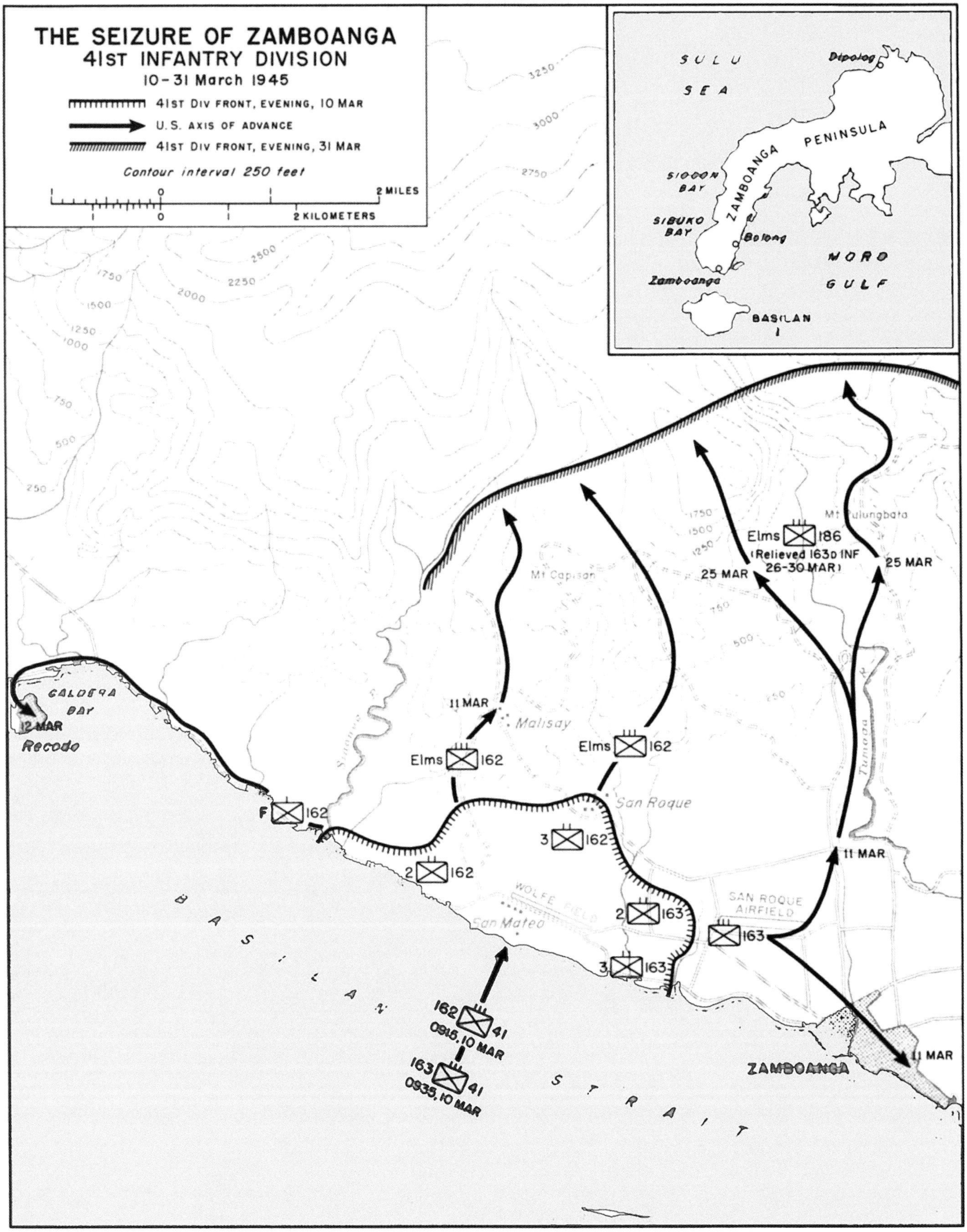

Zamboanga. *Courtesy of the Center for Military History.*

Left to right: Newell Whitt (pilot), H.C. O'Brien (radar observer), and two ground crewmen. The four men are standing in front of 550th NFS P-61 Black Widow 42-39576 *OHLAMI* based on the initials of the air crew's home states (Ohio, Louisiana, and Michigan). Note the survival knife on Whitt's waist and the shoulder holster with pistol carried by O'Brien. *Courtesy of AFHRA.*

weeks, resulting in 220 American army soldiers killed in combat and approximately 6,400 Japanese deaths.[35]

Marine Corps aircraft of Marine Air Group (MAG) 12 arrived at Zamboanga on 15 March, followed by MAG 35 a week later. Detachment A of the 419th NFS flew into what would become Moret Air Field at Zamboanga the day after the arrival of MAG 12. Living conditions at Zamboanga were a far cry from life on Morotai. Detachment A personnel fondly remembered the fresh meat and butter they received almost daily on Morotai. At Zamboanga the men lived for much of March on field rations. Beer rations on Morotai totaled 24 bottles per man; the rations at Zamboanga for March fell to 10 bottles per man. Considerable efforts were made to secure lumber from nearby islands for the construction of unit buildings. Operations Sergeant Jesse Gardner remembered all of the Japanese equipment that littered the area:

"There are a lot of Japanese pillboxes around and a few booby traps. Right beside our area are a couple of Japanese Bettys, which had apparently been bombed in their revetments."[36]

Detachment A flew 71 combat missions in March, with 50 of these from Zamboanga. The air operations from Zamboanga included 17 dusk or dawn patrols, 22 night patrols, two scrambles, and nine convoy coverage missions. During the night of 20 March Captain Don Dessert (Detachment A Commander) and First Lieutenant James Schroth were scrambled in response to Japanese artillery that was shelling Allied naval vessels eight miles west of Zamboanga. The pilots made passes over the general location of the Japanese gun positions, forcing the enemy to cease firing out of concern of disclosing their exact positions from the muzzle flashes. While

Moret airfield on Zamboanga, Philippines. The sign is a play on words. A song written sometime in the early 20th century and sung by many Americans stationed around Zamboanga during World War II opened with the line "Oh, the monkeys have no tails in Zamboanga." *Courtesy of Don Dessert.*

419th NFS Enlisted Mess Hall with chow line. *Courtesy of Don Dessert.*

419th NFS P-61 Black Widow 42-5563 #324 *Old Salty Dog II*. Note the protective cover over the canopy areas of the aircraft. Pilots and ground crews paid close attention to the condition of the plexiglas canopy of the Black Widow's cockpit. Dirt and scratches could block or distort any small light emitted by a bogey under pursuit and prevent the plane from being spotted or tracked. The maintenance personnel noted that fine particles of coral dust were very abrasive and could easily scratch the surface. Canopies were often covered to protect the plexiglas from the coral unless the P-61 sat on night alert. *Courtesy of Don Dessert.*

419th NFS P-61 Black Widows. 42-39473 #330 *Night Vision* is in the center of the photo with canvas over the canopy. *Courtesy of Don Dessert.*

not exciting, this type of mission served a useful purpose to halt the shelling; however, Japanese manned artillery remained concealed less than 10 miles from the airfield.

April proved to be a very busy flying month, with the Detachment amounting a total of 224 combat missions, including dusk and dawn patrols over Zamboanga, the Malabang beachhead, Jolo beachhead, and the Sanga Sanga beachhead. Although the Detachment flew a heavy mission load, necessitating the temporary transfer of two P-61s from Palawan, the Black Widows did not encounter any Japanese aircraft during the month.

On the morning of 1 April First Lieutenant Guy Scheetz flew his P-61 in support of efforts to spot Japanese artillery positions shelling the airfield at Zamboanga. Operations Sergeant Jesse Gardner remembered the enemy barrage that morning:

"We have no foxholes on the line or in the area, so we hugged the ground behind some boxes. The shells came closer all the time, so we took off for a searchlight position where they had a few sandbags. [They] dropped about forty shells altogether but didn't damage anything."[37]

Whenever the Japanese fired their weapons Scheetz would dive from his patrol altitude in an effort to spot them. In response, the enemy guns would cease operations. Scheetz also provided spotter services for the friendly guns firing counter battery missions against the Japanese guns. Scheetz would spot where the American artillery rounds landed and radio the results back to the gun crews.[38]

Japanese night fighters managed one raid on the American airfield at Zamboanga in April. In the pre-dawn hours of 2 April a Japanese single engine aircraft [probably a Val] managed to fly undetected by American radar to Zamboanga and bomb the airfield and escape before the Detachment A airborne night patrol could be contacted and directed back to hunt the raider. Sergeant Jesse Gardner recorded the attack in his diary:

"He dropped two five-hundred pound bombs beside the [air] strip, the second about one hundred feet from our line. No warning was sounded till [sic] after he'd dropped his bombs and left. Big chunks of mud fell a few feet from where [Sergeant] John [Godsey] was sleeping on alert."[39]

Eight days later, on 10 April Detachment A lost a P-61 and its crew during a combat mission over Jolo Island. The Detachment Black Widows were providing night fighter coverage over the American landings on Jolo Island, located along the chain of islands between Zamboanga and Borneo. At 5:30 AM pilot Second Lieutenant David Amerman, along with his R/O Second Lieutenant Warren Bakewell, departed Moret Field for the journey to Jolo, where they would commence a dawn patrol over the invasion fleet and beachhead. While on patrol over the beachhead area Amerman spotted what

Detachment A Commander Captain Don Dessert. Dessert and Intelligence Officer First Lieutenant John Guzey formed a team to investigate the combat related crash of Second Lieutenant David Amerman (pilot) and Second Lieutenant Warren Bakewell (radar observer) near Jolo, in the Philippines. It was determined that Amerman had inadvertently struck a coconut tree during an attacking pass on a Japanese vehicle. While the official crash report is silent about Japanese casualties in the attack, the investigation team confirmed that Amerman destroyed a Japanese staff car and two trucks with 20mm cannon fire during the attack. *Courtesy of Don Dessert.*

appeared to be a Japanese staff car and a couple trucks traveling along a road. He radioed the local Fighter Control, which was based on a destroyer lying off the island. Fighter control confirmed that there should not be an American vehicle in the vicinity reported by Amerman and cleared the P-61 pilot to engage the target. The P-61 made several low altitude strafing passes on the vehicles with 20mm cannon. On his last pass Amerman flew too low and a wing of the Black Widow clipped a coconut tree. The impact sheared off a section of the wing, forcing the P-61 into a dive which the pilot could not correct due to his low altitude. The Black Widow exploded, strewing wreckage over a 2,000 yard area southwest of Jolo town.

News of the crash shocked everyone in Detachment A. The 419th NFS had not lost an air crew member to a crash since Captain Howard Daniel in October 1944. The facts behind the incident were confused, and rumors spread quickly among the personnel. The initial information passing between the Detachment members noted that Amerman had been shot down by Japanese anti-aircraft fire, and this story persisted for days until the facts became sorted from the fiction. According to Sergeant Gardner, he heard that an American PT boat observed anti-aircraft fire striking the P-61, forcing it to crash.[40]

Detachment A Commander Captain Don Dessert and Intelligence Officer First Lieutenant John Guzey formed an investigating team to determine exactly what happened to Amerman and Bakewell that morning over Jolo. Following interviews and an investigation of the crash site, it was determined that the pilot had inadvertently struck the coconut tree during an attacking pass on the Japanese vehicle and had not been struck by anti-aircraft fire. While the official crash report is silent about Japanese casualties in the attack, Sergeant Gardner noted in his diary that the investigation team confirmed that Amerman destroyed the staff car and two trucks with 20mm cannon fire during the attack.[41]

Conditions at Zamboanga improved for the personnel assigned there during April. Fresh food began arriving and, to the delight of all, fresh baked bread appeared on the menu every other day, making up for the slight decrease in beer rations to nine bottles for the month. The Detachment aerial gunners, under the direction of Staff Sergeant John Graham, constructed an enlisted men's day room. Four Black Widows from the 550th NFS flew into Zamboanga on 27 April to support the continuing Allied operations to the southwest of the peninsula and returned to Talcloban on 11 May.

An unidentified Army nurse and Captain Dick Stewart attending a party thrown by the 419th NFS officers in the Philippines. Note that Stewart, a 419th NFS pilot, is wearing the insignia of a medical doctor on his collar. 419th NFS vets admitted years later that the Army nurses could only depart their quarters area with medical personnel, so Stewart improvised and borrowed a set of medical insignia to help them get to the party. *Courtesy of Don Dessert.*

Officers and enlisted soldiers of the 419th NFS attended elementary Spanish language classes to assist them with conversing with the local Filipinos, who often welcomed the Americans into their homes for parties and other social gatherings. Sometimes these social meetings led to interesting cultural misunderstandings. Sergeant Gardner remembered one of the most infamous among 419th NFS members, although he would not name the hapless man. Two of the Detachment personnel were visiting a Filipino family on a very regular basis. One evening, one of the two men stopped by the family home and asked in particular to visit with one of their daughters. They talked for a few minutes, and before leaving he commented to her parents that she was a very nice girl. The family mentioned that she would wait for him. The next evening he returned to visit the family and again talked with the young girl. During the evening, the entire family lined up and congratulated him on his engagement to the girl. A party with considerable alcohol followed, and the soldier

A military chapel in the Philippines. Note the open sides covered with only a strip of canvas due to the heat. *Courtesy of Don Dessert.*

forgot much of what led to the celebration. The next night he returned, and the family asked when he intended to marry the girl. The terrified soldier finally realized the predicament he had inadvertently stumbled into and volunteered to travel to the Squadron headquarters on Palawan the next day. He informed his buddies that he probably would not return.[42]

Only 69 missions were flown by the 419th NFS from Zamboanga during June. The war continued to move away from the peninsula, and most of the Detachment's aircraft were temporarily transferred to help establish Detachment B on Sanga Sanga during the month. Detachment A flew with two or three P-61s during June. Detachment A performed 34 dawn and dusk patrols over Malabang, 17 convoy coverage missions, and one night scramble, which was caused by a friendly aircraft. Morale continued to be high among Detachment personnel. Food quality continued to improve, as fresh meat and butter appeared regularly at the mess hall. The Squadron C-47 delivered fresh fruit and vegetables acquired in Australia. Trading remained one of the primary means of acquiring comfort items for Detachment personnel. Captain Dessert mentioned that the Detachment "loaned" a P-38 to the Marines in exchange for a walk-in refrigerator. According to Dessert, "We had the best mess hall in the Pacific."[43] Cigarette rations were low, but a case of beer was available for each man in June. According to Intelligence Officer John Guzey, the available beer for the month "did much to uplift the morale of all."[44] Japanese aircraft did not approach Zamboanga, and the air crews flying with Detachment A did not encounter any enemy intruders during the month.

Detachment A flew fewer combat missions (only 16) during the month of July due to the smaller number of aircraft operating from Zamboanga and receipt of an order to move to Palawan before the end of the month. Between 1 and 10 July the air crews flew ten dusk and dawn patrols and six night patrols from Moret Field. The Detachment lost another P-61 on 10 July. Captain Bernard Knott (pilot) and Flight Officer Donald Huntoon (R/O) were transporting three passengers, First Lieutenant William Rees and First Lieutenant Marion Flansberg, both 419th NFS pilots, and their dog from Zamboanga to Jolo Island. The men departed Moret Field at 11:15 AM and were nearing Jolo when the port engine died, followed later by the starboard engine. Knott ditched the Black Widow in the sea about 20 miles northeast of Jolo. The four men and the Detachment mascot managed to scramble safely from the P-61, which sank within 40 seconds of settling in the water. The crew floated in two life rafts for only 20 minutes before spotting two Filipino outrigger sailboats approaching. The Moro sailors rescued the P-61 crewmen and transported them to a U.S. Navy PT boat base on Jolo the next morning.

A PT boat carried the crew back to Zamboanga. Due to the loss of the Black Widow in the sea maintenance personnel were not able to determine the cause of failure for the engines. Detachment A personnel were simply relieved that all four men were safe.[45]

Life at Zamboanga continued to improve for Detachment A. Fresh meat, butter, and eggs were regularly on the menu at the mess hall, and the area did not experience any Japanese night intruders during the month. Although Detachment B lost an aircrew man to an accident at Sanga Sanga, Detachment A was relieved at the safe return of their four personnel following the crash near Jolo. Cigarettes were in short supply during the month, but the 24 bottle issue of beer for each man in July nearly made up for this inconvenience. Following the completion of combat operations within flying distances of Zamboanga, Detachment A was ordered to prepare to move and join the Headquarters Detachment on Palawan. Operational missions from Zamboanga ceased on 11 July, the day after Knott crashed near Jolo. Movement by Curtiss-Wright C-46 Commando and C-47 cargo planes and a U.S. Navy vessel commenced on 18 July and was completed by 26 July. The transfer to Palawan took longer than expected due to both of the Detachment's P-61s being down for maintenance. Upon arrival at Palawan, Detachment A was joined by the personnel from Detachment B on Sanga Sanga and all reveled as the 419th NFS became a single unit again.

Sanga Sanga (June 17 – 31 July 1945)

Black Widows of the 550th NFS arrived at Sanga Sanga on 17 June. They were joined by Detachment B of the 419th NFS on 20 June. Sanga Sanga lies southwest of Jolo Island and just east of Borneo. The mission of the 13th Air Force's night fighters included fighter coverage of Sulu Island, as well as the Sulu Archipelago, stretching between Mindanao Island of the Philippines and Borneo Island in the Dutch East Indies. 419th NFS Commander Lieutenant Colonel Norman Jacob formed Detachment

A rare collection of at least eight P-61 Black Widows from up to three squadrons (419th NFS, 550th NFS, and 418th NFS of the 5th Air Force) on the airstrip at Sanga Sanga. In the immediate foreground is 419th NFS P-61 Black Widow 42-39450 #329, and behind it is another 419th NFS Black Widow, 43-8238 #332 *Vivacious Vivian*. *Courtesy of Don Dessert.*

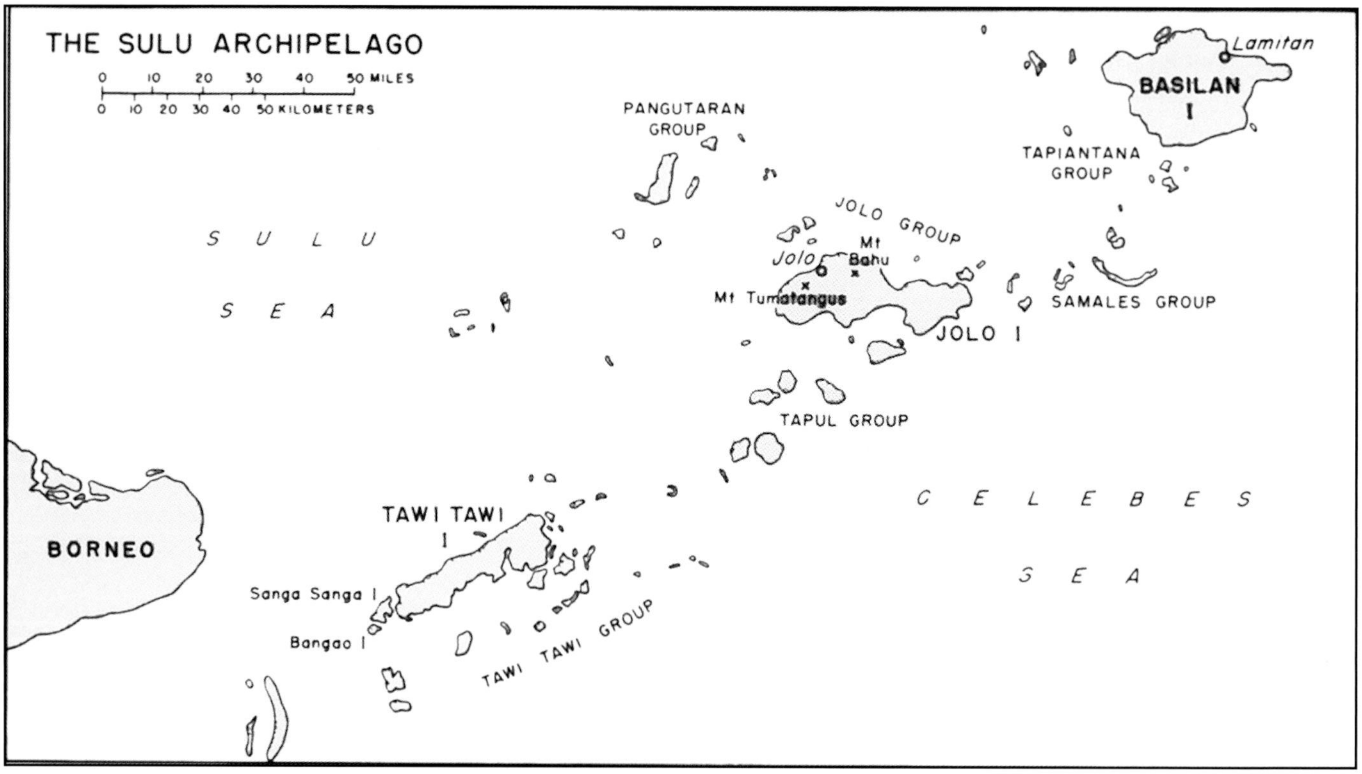

Sanga Sanga (located adjacent to Tawi Tawi). *Courtesy of the Center for Military History.*

B by transferring and combining planes and personnel from the Headquarters Detachment on Palawan and Detachment A at Zamboanga, and the first 419th NFS P-61s arrived on 20 June. A second group of 550th NFS Black Widows from Palawan joined their squadron mates at Sanga Sanga on the same day as the arrival of the 419th NFS. The 550th NFS Black Widows formed the squadron's Detachment A.

The Black Widows on Sanga Sanga formed a temporary organization consisting of assets from the 419th NFS (seven P-61s), 418th NFS of the 5th Air Force (two P-61s), and the 550th NFS (eight P-61s). Each squadron provided an operations officer who met collectively and cooperated with the other two officers in setting flying schedules for the organization. During June, aircraft of Detachment B (419th NFS) flew 46 combat missions and Detachment A (550th NFS) made 45 combat missions from Sanga Sanga.

A mechanical problem on the night of 25 June prevented an intercept of Japanese aircraft attacking American warships. The Detachment B (419th NFS) P-61 could not take off due to the mechanical problem and there were not any replacement aircraft available to fly the mission to cover the warships. That night, between five and seven Japanese aircraft carrying torpedoes attacked American minesweepers in the Balikpapan area. The presence of a P-61 night fighter could have led to a spectacular air battle with the intruders.

First Lieutenant John McGhee (without cap). *Courtesy of Leslee Lukas Mays.*

First Lieutenant Harold Michels, the 419th NFS pilot who shot down the Japanese Kawasaki Ki-45 Nick on 13 June, was now flying with Detachment B. He was scrambled at the end of the month in response to a radar contact approaching Sanga Sanga. Michels' R/O acquired and lost radar contact on the bogey three separate times. The crew returned to Sanga Sanga without finding the unknown aircraft.

Detachment B (419th NFS) lost a P-61 on 20 June, the first day 419th NFS Black Widows arrived on Sanga Sanga. The runway on Sanga Sanga was constructed of coral, which becomes very slippery when wet. First Lieutenant John McGhee, appointed commander of Detachment B, landed his P-61 after a rain and found his plane skidding across the wet airfield. Damage to the airplane was so extensive that the Black Widow was relegated to serve as a "Hangar Queen" to be salvaged for spare parts for flyable aircraft. The crew walked away from the accident.[46]

Detachment A (550th NFS) pilots made 142 combat flights, while Detachment B (419th NFS) flew 63 combat missions from Sanga Sanga during July. The 13th Air Force night fighters lost three P-61s at Sanga Sanga in July, including one on 4 July that resulted in the death of a 419th NFS air crew member. Air operations were cancelled on 3 July due to heavy rain. At 4:40 AM the next morning a P-61 with First Lieutenant Robert Laubach (pilot), Second Lieutenant Loren Kane (R/O), and Sergeant Ettore Fiore (gunner) attempted to take off to cover a convoy. The Black Widow became airborne, but crashed back onto the air strip seconds later and burned. Laubach and Fiore managed to get out of the aircraft with only minor injuries, but rescue personnel found Kane dead and lying in a foot of water behind the P-61. Medical personnel disclosed that Kane died from pulmonary asphyxiation caused by the fumes released by the burning fuel. Second Lieutenant Clarence "Clem" Green, an R/O, remembered the accident that night. First Lieutenant Michels was scheduled to also fly a mission on 4 July and needed an R/O to accompany him. Michels awakened Green and explained that he needed a replacement R/O for the upcoming dawn patrol:

"We didn't go because our conversation was interrupted by the noise of a P-61 taking off, followed by a flash of light and a thunderous boom. Neither of us had much of an appetite for flying after that incident."[47]

A potential accident was averted four days later on 8 July during an operation to cover a convoy. A 419th NFS P-61 piloted by First Lieutenant Charles Metzler lost an engine while over the convoy 230

Left to right: Second Lieutenant Loren Kane (radar observer), Sergeant Leonard Sherman (gunner), and First Lieutenant Robert Laubach (pilot). Laubach and Kane, along with Sergeant Ettore Fiore (gunner), attempted to take off to cover a convoy in June 1945. The Black Widow became airborne, but crashed back onto the airstrip seconds later and burned. Laubach and Fiore managed to get out of the aircraft with only minor injuries, but rescue personnel found Kane dead and lying in a foot of water behind the P-61. Medical personnel disclosed that Kane died from pulmonary asphyxiation caused by the fumes released by the burning fuel. *Courtesy of AFHRA.*

419th NFS Pilot First Lieutenant Robert Laubach (right) flew the P-61 Black Widow that crashed upon take off at Sanga Sanga in July 1945. *Courtesy of AFHRA.*

miles southeast of Sanga Sanga at an altitude of 7,500 feet. Metzler turned his Black Widow toward Sanga Sanga for an emergency landing. Although he continued to lose altitude during the return flight, he managed to safely land the plane. The next day the 550th NFS lost a P-61 in an accident. First Lieutenant James Colby took off at 3:57 AM for a dawn patrol over the Balikpapan area. As the Black Widow lifted off the pilot experienced trouble, resulting in a crash 600 yards from the air strip.[48]

Morale among the night fighter personnel on Sanga Sanga was high from June through July despite the lack of fresh food, poor living conditions, and typhoons. A C-47 from the 419th NFS Headquarters Detachment on Palawan arrived daily to shuttle mail and fresh food for their personnel. The two aircraft of the 418th NFS departed Sanga Sanga on 9 July. News arrived on 20 July that Detachment B of the 419th NFS should cease all operational flying and rejoin the other elements of the squadron at Palawan. Personnel made the arrangements for the move and departed Sanga Sanga on 27 July. The P-61s of the 550th NFS returned to Tacloban on 31 July. World War II would end less than two weeks later with the surrender of Japan.

A shower point on Sanga Sanga. Compare this shower with the more elaborate one on Middleburg Island and pictured in Chapter 5. *Courtesy of Don Dessert.*

A latrine utilized by the night fighters on Sanga Sanga. *Courtesy of Don Dessert.*

Night fighter living quarters on Sanga Sanga. *Courtesy of Don Dessert.*

Lining up at the chow tent on Sanga Sanga. *Courtesy of Don Dessert.*

8

End of the War and Deactivation of the 13th Air Force Night Fighters

Don't Give me a P-61, for night flying is no fun.
They say it's a lark, but I'm scared of the dark.
Don't Give me a P-61.

Last stanza of "Make Me Operations,"
a WWII era USAAF song
Composer Unknown

While personnel rotations of the 550th Night Fighter Squadron (NFS) were fairly stable due to the short period the unit had been in combat, April and May 1945 proved to be a major turning point in the history of the 419th NFS, as many of the surviving original air crew members rotated back to the United States. Captain Don Dessert departed for the United States on 18 April after completing his combat tour as a pilot in the theater and was replaced by First Lieutenant James Schroth as Detachment A commander. Staff Sergeant Frank McCormack, Dessert's gunner during the October 2, 1944, confirmed kill of a Japanese Mitsubishi Ki-46 Dinah near Middleburg Island, also departed for the United States. Other original 419th NFS members who rotated back to the United States included Captain Al Lukas, who had recovered from his severe bout of malaria and returned to the Squadron in the Philippines just prior to departing the theater. Lukas bagged the Squadron's first kill, a Mitsubishi Ki-21 Sally bomber, while flying from Noemfoor. Captain Dick Stewart left the theater with the original intention to complete leave and return to command the Squadron. Stewart bagged a Japanese Mitsubishi G4M Betty bomber as the 419th NFS' only P-38 kill, and also had an unconfirmed kill of the same type of aircraft. Stewart would never return to be the Squadron commander due to the ending of the war at the end of the summer of 1945. Captain Robert Oates, another pilot, joined the departing group.

Two original Radar Observers (R/Os) left the theater in April. One of these R/Os was First Lieutenant James Allen, who had flown with Captain Dick Stewart when they had spared the American B-24 bomber that lacked a functioning identification signal. Allen would sail with Stewart on the same troop ship back to the United States and meet that B-24 crew for dinner in San Francisco. The other R/O was First Lieutenant John Radzieta, who had flown with many pilots while overseas. Returning gunners included Technical Sergeant Glen DeForrest, who had flown with Lukas and Schroth during the nights each pilot scored a confirmed kill. DeForrest, with two

419th NFS Pilot First Lieutenant Herbert Johnson. *Courtesy of AFHRA.*

confirmed kills as a gunner, ended the war as the 419th NFS' highest scoring air crew member. In May, pilots First Lieutenant Guy Scheetz, First Lieutenant Kenneth Voy, and First Lieutenant George White departed the theater, along with R/Os First Lieutenant Paul Ferguson, First Lieutenant William Gillette, First Lieutenant Frederic Kahn, First Lieutenant Ralph Lillard, and First Lieutenant Burton Tracy. The air crew members who had transferred into the 419th NFS from Detachment B 6th NFS on Guadalcanal had already departed the theater. By the end of May, the air crews of the 419th NFS consisted primarily of men who had arrived as replacements after the Squadron had entered the Pacific Theater.

As August 1945 opened, American troops and Filipino guerillas continued to clear the Philippines of Japanese forces who had withdrawn into the hills and jungles throughout the islands. At the same time, General Douglas MacArthur initiated plans to stage 22 divisions in the Philippines to support the upcoming invasion of Japan, scheduled for November 1945. The reassembling of the 419th NFS detachments into a single airfield on Palawan and those of the 550th NFS at Tacloban were important elements in the plan to subdue Japan. 419th NFS air and ground crews prepared to mark the unit's second anniversary in the theater by providing night fighter coverage for the invasion of the Japanese home islands; the 550th NFS received orders to transfer to Okinawa and cover the same operation.

The 419th NFS continued to provide night fighter defense at Palawan from 1 to 15 August. The squadron flew eleven dawn and dusk patrols, one night patrol, two search missions, and one escort mission during this two week period. The last combat mission of a 13th Air Force night fighter occurred on 8 August when a 419th NFS P-61 intercepted an American Navy PB4Y-2 Privateer with a failed Identification Friend or Foe (IFF) transponder. The Black Widow identified the aircraft as friendly and escorted it to Palawan. The 419th NFS prepared to move to Leyte as the 13th Army Air Force readied its units for the invasion of Japan. However, with the announcement of Japan's surrender the movement orders were cancelled, and the 419th NFS personnel remained in place on Palawan. The men turned to activities to keep them occupied, including educational courses and sports. Daily work required within the squadron was primarily accomplished by the recent arrivals, as the veterans turned to some well deserved time off. With the end of the war came a decreased quality in the Quartermaster food supplies: fresh meat became scarce; flour disappeared, resulting in a closing of the bakery; and fresh vegetables vanished. The unit made two "Fat Cat" flights in the squadron C-47 to Australia to acquire food to supplement the canned and dehydrated items available through the normal Army supply channels. Post exchange items decreased, and even the beer ration was cut to 18 bottles per man for the month. Most of the veteran air crew members and all of the Ground Control Intercept (GCI) operators returned to the United States while new air crews continued to arrive and join the 419th NFS. Over the next couple of months the veterans of the 419th NFS returned home and the squadron, with newly arrived air and ground crews, eventually moved to Clark Air Field.[1] The 419th NFS remained in the Philippines until deactivation in 1947.

The mission of the 550th NFS for August 1945 consisted of night rather than day defense of Leyte with the return of the squadron's P-61s. Each night the squadron maintained two Black Widows on strip alert between 7:00 PM and 7:00 AM. The 550th NFS did not scramble any fighters during the month. The major activity involved making the preparations to transfer the squadron from Leyte to Okinawa on 13 August as Allied forces moved closer to the invasion of Japan.

419th NFS Radar Observer Henry Bettier. Note the unofficial 419th NFS patch on his leather jacket. *Courtesy of AFHRA.*

The Mid-Atlantic Air Museum launched an ambitious program to recover the 550th NFS P-61 Black Widow that crashed on Mount Cyclops, in New Guinea. Only three other P-61s are known to be in existence. One is in the People's Republic of China; one is in the National Museum of the United States Air Force in Dayton, OH; and the third is at the National Air and Space Museum - Steven F. Udvar-Hazy Center, located adjacent to Dulles International Airport near Washington, D.C. Rumors persist of two additional P-61s in storage crates located in the People's Republic of China. *Courtesy of the Mid-Atlantic Air Museum.*

With the receipt of the news on 10 August of Japan's surrender the transfer to Okinawa was cancelled. Jubilation was followed by activities to keep the men busy. Flight operations ceased on 17 September, and the Black Widows were soon turned over to a depot at Clark Field, near Manila. The squadron personnel departed the Pacific Theater and arrived in the United States on January 3, 1946. The Army deactivated the 550th NFS the following day.[2]

The limitations of the P-61 as a night fighter were becoming evident toward the end of World War II. The P-61s of the 421st NFS (5th Air Force) could not catch the Japanese Nakajima Ki-43 Oscar fighters that were prevalent in the night sky over Leyte during December 1944. United States Marine Corps Grumman F6F Hellcats eventually replaced the Black Widows in the hunt to catch the Oscars.[3] The Army Air Force sought a replacement for the P-61A and B aircraft, which had been designed in the early stages of the war. Experiments continued with a two-seat version of the P-38 Lightning that had proven itself as a nimble fighter in day and night operations. The aircraft, known as the P-38M, carried an AN/APS-4 radar pod system under the nose and an R/O who sat elevated and behind the pilot. Although at least four P-38Ms were dispatched to the Philippines, the plane was never flown in combat prior to the end of the war. After Japan's surrender the P-38M was fielded in at least one night fighter squadron, but did not replace the P-61 Black Widow. The Army Air Force did produce

The Mid-Atlantic Air Museum unveiled the partial restoration of the 550th NFS' P-61 Black Widow during one of the final gatherings of the World War II Night Fighters' Association (veterans have disbanded the organization). The Museum plans to restore the Black Widow to flying status. *Courtesy of the Mid-Atlantic Air Museum.*

small numbers of a new P-61 version. The new model, the P-61C, added increased speed and a higher operational ceiling, as well as "fighter brakes" on the wing panels to help increase the rate of turn. The war ended before the P-61C could be rolled out in large numbers, and the few planes that were completed were diverted for research and test missions. The F-82 Twin Mustang, a twin fuselage version of the North American P-51 Mustang, replaced the P-61 Black Widow beginning in 1948.[4]

There has been some criticism of the World War II night fighter program. An examination of the 13th Air Force night fighters shows that Detachment B of the 6th NFS lost at least five aircraft while scoring one confirmed and one probable kills against the Japanese between March and November 1943; twenty-eight aircraft of the 419th NFS were destroyed due to accidents or combat in exchange for six confirmed and one probable kills of Japanese planes between November 1943 and August 1945; and the 550th NFS tallied one destroyed Japanese fighter on the ground compared to losing at least two of their own planes between January 1945 and August 1945. Such criticism is unfair, and demonstrates a lack of understanding of the situation faced by the 13th Air Force night fighters. The night fighters faced the same hazards of their counterparts in day fighters, except they had the disadvantage of doing everything at night. America's night fighter pilots suffered a higher percentage of combat deaths than their day counterparts during the war.

There are three reasons for the inability of the 13th Air Force night fighters to make more kills than they did between 1943 and 1945. First, they had the disadvantage of flying the P-38 Lighting on searchlight missions and the P-70 Nighthawk from March 1943 to June 1944. Lacking onboard radar, the P-38 was forced to rely on spotting the exhaust flames or silhouette of a Japanese fighter unless it could be picked up by a ground based searchlight. The P-70, despite having its own radar system, was ineffective as a night fighter, lacking the speed and altitude to bag Japanese raiders. With the introduction of the P-61 Black Widow this situation changed. The night fighters finally had a plane that was capable of finding and downing Japanese aircraft, but there had to be enemy aircraft in the sky in order to make kills.

The same month the P-61s of the 419th NFS became operational, the 13th Air Force moved into a supporting role behind the 5th Air Force. The night

The North American F-82 Twin Mustang replaced the P-61 Black Widow as the Air Force's main night fighter from 1948-1951. The F-82 scored the first three American aerial victories of the Korean War. *Courtesy of Don Dessert.*

The 550th NFS radar section. The SCR-720 radar system dish on the table fits into the forward nose compartment of the P-61 Black Widow. The radar emitter, located below the pilot's cockpit, sent out the signal. When the signal bounced off an object it returned to be received by the dish in the nose. The SCR-720 system then interpreted the signal and produced a blip on the radar observer's scope from which he could determine distance and altitude of the target. *Courtesy of AFHRA.*

fighter squadrons of the 5th Air Force defended the forward edge of General MacArthur's offensive, while those of the 13th Air Force tended to remain in areas experiencing less combat. A few surprising figures clearly illustrate this subordinate role played by the 13th Air Force. Captain Lukas scored the first P-61 kill of the 13th Air Force in August 1944. This was the *only* 13th Air Force kill (day and night) in the month of August 1944. The combat scores for 1945 are even more illuminating. The 419th NFS destroyed two Japanese aircraft during the eight months of combat in 1945. At first glance one might question the combat effectiveness of the unit. However, these two kills are seen in a new light when compared to the fact that all *eight* fighter squadrons of the 13th Air Force downed only 22 Japanese planes from January 1945 to August 1945. Between May and

The emitter dish of the SCR-720 radar system carried in the nose of the P-61 Black Widow. *Courtesy of Don Dessert.*

August 1945 the 13th Air Force fighter squadrons shot down a total of four Japanese aircraft—one of these bagged by the 419th NFS.[5] The reduction in Japanese night raiders encountered by the 13th Air Force was a primary reason why its night fighters began flying more daylight harassing missions in the Philippines.

Being in a subordinate role behind the 5th Air Force meant the fighter squadrons of the 13th Air Force faced fewer Japanese aircraft. Author D. Clayton James was able to interview Lieutenant General Earl Barnes and Major General St. Clair Streett about the relationship of the 5th and 13th Air Forces for his two volume work entitled *The Years of MacArthur*. Barnes commanded the 13th Fighter Command in 1944 and assumed command of the 13th Air Force late in the war during 1945. According to James, Barnes informed him that "[Lieutenant General Ennis P.] Whitehead [commander of the 5th Air Force after June 1944] was 'caustic' toward the 13th's officers and was 'constantly sniping at us." Major General Streett commanded the 13th Air Force from June 1944 when it was presented with a subordinate role to the 5th Air Force. James wrote:

> "In Streett's outfit there was 'some chaffing on the part of the staff' about their missions, and in January 1945, Streett disgustedly proposed that the 13th be deactivated and its air units absorbed into the 5th."[6]

Despite the assurances of General MacArthur and Lieutenant General Kenney that the 13th Air Force would play a more prominent role in the offensive operations, the situation did not change.

Thirteenth Air Force discontent with its role in the air war was not unique. Fighter pilots with the Royal Australian Air Force expressed dismay at being left behind on Morotai for ground attack missions against bypassed Japanese garrisons as American squadrons moved forward to the Philippines. Many senior Australian pilots presented their resignations in April 1945, prompting an investigation into their complaints. Although the situation changed little prior to the end of the war, the incident did serve to highlight general discontent among the growing number of Allied fighter pilots during a time in the war when there were fewer Japanese fighters in the Southwest Pacific.[7]

Third, night fighter defense in the 13th Air Force primarily fell on one unit prior to 1945 – the 419th NFS. The four 5th Air Force night fighter squadrons (Detachment A 6th NFS, 418th NFS, 421st NFS, and the 547th NFS) downed 37 Japanese aircraft between 1943 and 1945, while those of the 13th Air Force (Detachment B 6th NFS, 419th NFS, and the 550th NFS) bagged only seven confirmed Japanese planes. Detachment B 6th NFS departed the 13th Air Force in December 1943, and the 550th NFS did not become active for combat until January 1945. The 419th NFS carried the weight of the 13th Air Force's night fighter defense between November 1943 and January 1945 while flying P-61 Black Widows for only the last six months of this period.

The night fighters of the 13th Air Force downed seven confirmed kills while losing at least thirty-five aircraft due to accidents and combat. Was this unequal exchange worth the effort? Simply stated, yes. The mere presence of the night fighter squadrons reduced the number of Japanese intruders striking Allied bases after dusk. Considerable evidence exists of radar contacts abandoning their missions when realizing they had been discovered by American radar directing night fighters. Many other intruders were thwarted in their attempts to accurately bomb their targets as they sought to lose pursing American night fighters. It is impossible to calculate how many Allied lives may have been saved by fielding a credible night fighter defense force against the Japanese. In addition, the mere presence of the night fighters increased the morale of those on the ground and reduced the stress of dodging Japanese night intruders. The author has never met a former World War II night fighter air crew member who questioned the efforts of the program despite the high percentage of losses.

Night combat missions are now common, and often a preferred tactic when viewing Air Force operations against Iraq and Serbia. Combat pilots are routinely trained for night operations and carry equipment, including night vision devices that permit them to dominate the night skies. Looking back at World War II, one has to simply admire the courage of America's night fighters to help pioneer the art of night aerial combat.

Endnotes

Chapter 1

1. For more information on the development of aerial night fighting during World War I and the early years of World War II, see Den Delve, *Nightfighter: The Battle for the Night Skies*, London: Cassell and Co., 1995.

2. RAF flight log of PO Al Lukas and many conversations with Lukas, the father-in-law of the author.

3. Conversations with Al Lukas and other veterans of the 419th NFS; USAF Historical Division, *Development of Night Air Operations 1941-1952*, USAF Historical Study No. 92, Air University, 1952; Garry R. Pape and Ronald C. Harrison, *Queen of the Midnight Skies: The Story of America's Air Force Night Fighters*, Atglen, PA: Schiffer Publishing, 1992.

4. Flight Lieutenant Karel M. Kuttelwascher, interviewed by Captain Paul Aiken. "Psychological Effect of Night Fighter Intruder Tactics," AAFSAT *Intelligence Reports*, Air Forces School of Applied Tactics, September 1943.

5. RAF flight log of PO Al Lukas and many conversations with Lukas.

6. Don Dessert. Interview with the author. August 16, 2000.

7. Dick Stewart. Interview with the author. September 3, 2000.

8. Harold Young. Interview with the author. September 2, 2000.

9. Don Dessert. Interview with the author August 16, 2000.

10. Sandy Bauman. Interview with the author. February 1, 2000.

11. USAAF flight log of Al Lukas and many conversations with Lukas.

12. Harold Young. Interview with the author. September 2, 2000.

13. Robert O'Brien. Interview with the author. April 4, 2000. The A-24, essentially an Army version of the Navy's SBD Dauntless, served as a dive bomber early in the war. The plane had mixed reviews as a dive bomber for the Army, and many were allocated to squadrons as administrative and training planes. On at least one occasion, the 419th NFS utilized one of its A-24s to move a detachment's beer supply to its new base.

14. Sandy Bauman. Interview with the author. February 1, 2000.

15. John Radzieta, Interview with the author, February 27, 2004.

16. Sandy Bauman. Interview with the author. February 1, 2000.

Chapter 2

1. Details of the Allied operations in 1942 can be found in Wesley Frank Craven and James Lea Cate, editors. *The Army Air Forces in World War II: Plans and Early Operations, January 1939 to August 1942*. Chicago: University of Chicago Press, 1950. This seven volume series, published in the years after World War II, was prepared by incorporating historical studies written during the war from official military documents and correspondence under the guidance of the Air Force Historical Division. Craven was the Chief of the Operations History Branch of the Army Air Force Historical office, and Cate served as the historical officer of the 20th Air Force after June 1944. Despite the age of these works, they are still good summaries of Army Air Force operations during World War II and are, for all practical purposes, the official U.S. Air Force history of the war. Another good review of this period is the official U.S. Army's historical record of World War II. Other historical reviews of this period include Alan Schom, *The Eagle and The Rising Sun: The Japanese-American War 1941-1943*, New York: W. W. Norton and Company, 2004; H.P. Willmott, *The Barrier and the Javelin: Japanese and Allied Pacific Strategies, February to June 1942,*

Annapolis, MD: Naval Institute Press, 1983; and Winston Groom, *1942: The Year That Tried Men's Souls*, New York: Atlantic Monthly Press, 2005.

2. For more details on the Guadalcanal campaign, see Richard B. Frank, *Guadalcanal*, New York: Random House, 1990; Eric M. Hammel, *Guadalcanal: Starvation Island*, New York: Crown Publishers, 1987; Samuel B. Griffith, *The Battle for Guadalcanal*, Philadelphia: Lippincott, 1963; and Richard Tregaskis, *Guadalcanal Diary*, New York: Random House, 1943.

3. For an excellent, easy-to-understand discussion of the complex command structures in the Pacific Theater see Louis Morton, "Pacific Command: A Study in Interservice Relations" in *The Harmon Memorial Lectures in Military History, 1959-1987*. Edited by Harry R. Borowski. Washington, DC: Office of Air Force History, 1988, pp. 129-152.

4. William F. Halsey with J. Bryan, III. "Admiral Halsey Tells His Story." Part Six. *The Saturday Evening Post*. Vol. 220, Issue 3, July 19, 1947, p. 34. The series of articles published in *The Saturday Evening Post* were essentially an abridged version of Halsey's autobiography (co-written with J. Bryan, III) *Admiral Halsey's Story*. New York: McGraw Hill, 1947.

5. More details on COMAIRSOLS can be found in William Wolf, *13th Fighter Command in World War II: Air Combat over Guadalcanal and the Solomons*, Atglen, PA: Schiffer Military History, 2004.

6. Problems with Japanese night raiders are discussed in Army Air Force Historical Office, Assistant Chief of Air Staff Intelligence Historical Division. *Guadalcanal and the Origins of the Thirteenth Air Force*. Army Air Force Historical Studies No. 35. July 1944; Army Air Force Historical Office. *The Thirteenth Air Force: March-October 1943*. Army Air Forces Reference History No. 20, 1944; and United States Air Force, Historical Division of the Air University. *Development of Night Air Operations: 1941-1952*. U.S. Air Force Historical Study No. 92, 1953.

7. Jim Allen. Interview with the author. April 2, 2006. It should be noted that the author has not found a piece of solid information proving or disproving the sounds a beer bottle makes when falling from an aircraft.

8. Army Air Force, Assistant Chief of Air Staff Intelligence Historical Division. *Guadalcanal and the Origins of the Thirteenth Air Force*. Army Air Force Historical Studies No. 35. July 1944, p. 163, and United States Army Air Force, Assistant Chief of Air Staff Intelligence Historical Division. *The Thirteenth Air Force: March-October 1943*. Army Air Forces Reference History No. 20, 1944, p. 44.

9. Ibid. For more details on the P-70 Nighthawk, see Garry R. Pape and Ronald C. Harrison, *Queen of the Midnight Skies: The Story of America's Air Force Night Fighters*, Atglen, PA: Schiffer Publishing, 1992, and Harry Gann. *The Douglas A-20 (7A to Boston III)*. London: Profile Publications, 1971.

10. Many secondary sources are contradictory on who flew as the R/O on this mission with Bennett. Some list Raymond Mooney as the R/O, but others place Sergeant Edwin Tomlinson in the P-70 with Bennett on April 19, 1943. Official USAAF records credit Mooney in the cockpit, and he is the R/O granted credit for the kill, along with Bennett, for downing the Betty bomber. The USAAF officially credited the pilot, R/O, and gunner (for P-61 Black Widows) with each kill of an enemy aircraft with an airborne radar directed fighter, such as the P-70 or later the P-61. Tomlinson was an R/O with Detachment B on Guadalcanal and did complete combat missions in the P-70 aircraft. Mooney remained with the 6th NFS after many of the Detachment B personnel were absorbed into the 419th NFS and is credited for an additional three Betty bombers as the R/O on P-61 Black Widow missions. See Wesley P. Newton, Jr., Calvin F. Senning, et al. *USAF Credits for the Destruction of Enemy Aircraft, World War II*, United States Air Force Historical Research Agency Study No.85, 1978.

11. Ehlert quoted in John W. Lambert. *Pineapple Air Force*. St. Paul, MN: Phalanx Publishing, 1990, p. 47. Information on the first P-70 kill can be found in *Guadalcanal and the Origins of the Thirteenth Air Force*, p. 46; "The Nocturnal Nighthawk," *Air Enthusiast* No. 27. Date and author unknown; and United States Air Force, Historical Division of the Air University. *Development of Night Air Operations: 1941-1952*. U.S. Air Force Historical Study No. 92, 1953.

12. *The Thirteenth Air Force: March-October 1943*, p. 55.

13. Ibid, pp. 47-49.

14. Quoted in "The Nocturnal Nighthawk," p. 67. Information on the downing of this Sally bomber can also be found in United States Air Force, Historical Division of the Air University. *Development of Night Air Operations: 1941-1952*. U.S. Air Force Historical Study No. 92, 1953.

15. Ibid.

16. Detachment B, Night Fighter Squadron memorandum *Historical Record of Organization*. December 15, 1943.

17. "The Nocturnal Nighthawk," pp. 68-69; *Army Combat Intelligence Report*, September 17, 1943.

18. Mitsubishi F1M Pete and Nakajima E8N Dave aircraft were utilized by Japan in this role.

19. Quoted in "The Nocturnal Nighthawk," p. 59. Additional information on this type of night fighter ruse can be found in 13th Air Force, *Army Combat Intelligence Report*, September 17, 1943. The latter document refers to the scheme as a general failure to lure and destroy Japanese aircraft at night.

20. Detachment B, Night Fighter Squadron memorandum *Historical Record of Organization*. December 15, 1943.

21. Blake, Steve. "Historian's Corner," *P-38 National Association Newsletter*. July 6, 2003, page 6.

22. The Ventura normally carried a crew of six personnel, although the night fighter version carried only three—a pilot, R/O, and gunner. The plane could reach a maximum speed of just over 320 mph with a combat range of over 1,600 miles. For more on the Ventura bomber see Alan C. Carey. *PV Ventura / Harpoon Units of World War 2 Osprey Combat Aircraft# 34).* Botley, United Kingdom: Osprey Publishing, 2002, and Charles L. Scrivner and W.E. Scarborough. *Lockheed PV-1 Ventura in Action.* Carrollton, TX: Squadron/Signal Publications, Inc., 1981. See also Henry I. Shaw, Jr. and Major Douglas T. Kane. *History of U. S. Marine Corps Operations in World War II. Volume II. Isolation of Rabaul.* Washington, DC: Historical Branch, U.S. Marine Corps, 1963.

23. John M. Elliott. "Solomon Islands: The Isolation" *Naval Aviation News.* November-December 1993, pp. 28-29 and Robert Sherrod. *History of Marine Corps Aviation in World War II.* Washington DC: Combat Forces Press, 1952, pp. 158-162. For a discussion of U.S. Navy night fighter radar development and training see Alan William Slater. *Mickey and the Night Fighters.* Naval History. Vol. 17, Issue 5, October 2003, pp. 48-51.

24. Colonel Charles Quilter II and Captain John Chapin. *A History of Marine Fighter Attack Squadron 531.* Washington, DC: U.S. Marine Corps History and Museums Division. 2001, p. 11 and Samuel Eliot Morison. *Breaking the Bismarcks Barrier 22 July 1942 – 1 May 1944.* Boston: Little, Brown, and Company, 1950, p. 346.

25. RAF flight logs of Al Lukas and many conversations with the author.

26. Richard Stewart. Interview with the author. June 25, 1999.

27. Richard Pitman. Interview with the author. June 24, 1999.

28. Richard Stewart. Interview with the author. June 25, 1999.

29. Eddy Ballard. Interview with the author. June 24, 1999.

30. Sandy Bauman. Interview with the author. June 23, 1999.

31. Bob O'Brien. Interview with the author. June 24, 1999.

32. Don Dessert. Interview with the author, March 10, 2007.

33. Bob O'Brien. Interview with the author. June 24, 1999.

34. Bob O'Brien. Interview with the author. May 6, 2007.

35. 419th Night Fighter Squadron. *Mission Report.* December 10, 1943.

36. Richard Stewart. Interview with the author. June 25, 1999.

37. John Radzieta. Interview with the author. February 27, 2004.

38. Sandy Bauman. Interview with the author. June 23, 1999.

39. Richard Stewart. Interview with the author. June 25, 1999, and Don Dessert. Interview with author. March 10, 2007.

40. Richard Stewart. Interview with the author. June 25, 1999.

41. 419th Night Fighter Squadron. *Mission Report.* December 26, 1943, and USAAF Flight Logs of Al Lukas.

42. Richard Stewart. Interview with the author. June 15, 1999.

43. 419th NFS. *Historical Record of the 419th Night Fighter Squadron from Time of Activation.* January 1, 1944.

Chapter 3

1. For more information see W. F. Craven and J. L. Cate, editors. *The Army Air Force in World War II. The Pacific-Guadalcanal to Saipan. Volume IV (August 1942 to July 1944).* Chicago: University of Chicago Press, 1950.

2. Torokina air strip opened for limited air operations less than one month prior to the day when Lukas utilized it as a staging point for his raid on Rabaul. Torokina lay within 65 miles of five Japanese airfields during this period and could be a dangerous location for pilots attempting to fly in and out of Bougainville. For a good description of the work to develop and protect Torokina air strip see W. F. Craven and J. L. Cate, editors. *The Army Air Force in World War II. The Pacific-Guadalcanal to Saipan. Volume IV (August 1942 to July 1944).* Chicago: University of Chicago Press, 1950.

3. The Mitsubishi F1M Pete and Nakajima E8N Dave were Japanese naval float planes. Both were two-seated, bi-wing aircraft developed in the 1930s. They were utilized primarily for reconnaissance, although the Pete in particular performed night bombing and air defense missions in the area of the Solomon Islands.

4. Belief that the Japanese did not utilize radar during World War II is an ongoing myth. Although they were slower to develop radar systems for military purposes than the other belligerents, the Japanese employed fixed and mobile ground-based, naval, and airborne radar systems. For more information on the use of radar by the Japanese in World War II see Yasuzo Nakagawa. *Japanese Radar and Related Weapons of World War II.* Walnut Creek, CA: Aegean Press, 1998. The radar set strafed by Lukas and his crew was destroyed later in the campaign and the remains are still on the island.

5. Information on the Al Lukas night intruder raid on Rabaul comes primarily from three sources. The first represents many informal discussions with Lukas, followed by more formal interviews prior to his death in 1995. The second source is Volume II of Al Lukas' Flight Logs. These are two bound volumes covering his years with the RAF in Europe and the USAAF in the Pacific. Lukas meticulously maintained his logs and included considerable detail for each of his significant combat missions. The third source is a collection of official 419th Night Fighter Squadron documents prepared for submission to the 13th Air Force headquarters, including 419th NFS memorandum. *Final Mission Report*. Dated January 11, 1944, and 419th NFS memorandum. *Historical Report of the 419th NFS January 1 – January 31, 1944.* Dated February 5, 1944.
6. Don Dessert. Interview with the author June 24, 1999, and Dick Stewart. Interview with the author. June 25, 1999.
7. Masatake Okumiya and Jiro Horikoshi with Martin Caidin. *Zero!* New York: Dutton, 1956, p. 285.
8. 419th Night Fighter Squadron. *Mission Report*. Dated January 10, 1944.
9. 419th Night Fighter Squadron. *Mission Report*. Dated January 13, 1944, and Don Dessert, Interview with the author, April 22, 2007.
10. 419th Night Fighter Squadron. *Mission Report*. Dated January 15, 1944.
11. John Radzieta, Interview with the author, February 27, 2004.
12. Roger Egan letter to John Radzieta. No date.
13. James Allen. Interview with the author. September 6, 2002.
14. James Allen. Interview with the author. March 18, 2007.
15. Robert O'Brien. Interview with the author. June 24, 1999.
16. Jesse Henry Gardner. *Beachheads and Black Widows: A South Pacific Diary*. Pittsburgh, PA: Hermann Printing and Litho, Inc., 1995, p. 39.
17. Jim Allen. Interview with the author. May 17, 2007.
18. John Radzieta, Interview with the author, February 27, 2004; Don Dessert, Interview with the author, June 23, 1999; Sandy Bauman, Interview with the author, June 23, 1999; and Dick Stewart, Interview with the author, June 25, 1999.
19. Robert O'Brien. Interview with the author. June 24, 1999.
20. Richard Stewart. Interview with the author. June 23, 1999.
21. Don Dessert, Interview with the author, July 29, 2002.
22. Masatake Okumiya and Jiro Horikoshi with Martin Caidin. *Zero!* New York: Dutton, 1956, pp. 309-310.
23. James Allen. Interview with the author. June 24, 1999.
24. Richard Nixon. *The Memoirs of Richard Nixon*. NY: Grosset and Dunlap, 1978, pp. 28-29.
25. Richard Stewart. Interview with the author. June 25, 1999.
26. Ibid.
27. Ibid.
28. Sandy Bauman. Interview with the author. June 24, 1999.
29. Cleve Wilborn. Interview with the author. June 24, 1999.
30. Ibid.
31. Ibid.
32. Dick Stewart. Interview with the author. June 25, 1999.
33. Cleve Wilborn. Interview with the author. June 24, 1999.
34. Jesse Henry Gardner. *Beachheads and Black Widows: A South Pacific Diary*. Pittsburgh, PA: Hermann Printing and Litho, Inc., 1995, p. 43-50.
35. Ibid.
36. Don Dessert. Interview with the author June 23, 1999.
37. Don Dessert. Interview with the author June 23, 1999, and Dick Stewart. Interview with the author. June 25, 1999.
38. 419th Night Fighter Squadron. *Mission Report*. Dated May 4, 1944.
39. 419th Night Fighter Squadron. *Mission Report*. Dated May 13, 1944.
40. 419th Night Fighter Squadron. *Mission Report*. Dated May 6, 1944, and COMAIRSOPAC Report dated April 22, 1944.

41. 419th Night Fighter Squadron. *Mission Report*. Dated May 8, 1944.

42. For further information on the P-61 Black Widow, see Garry R. Pape and Ronald C. Harrison, *Queen of the Midnight Skies: The Story of America's Air Force Night Fighters*, Atglen, PA: Schiffer Publishing, 1992; Warren E. Thompson, *Northrop P-61 Black Widow*, Warbird Tech Series Volume 15, North Branch, MN: Specialty Press Publishers, 1997; Larry Davis and Dave Menard, *P-61 Black Widow in Action*, Aircraft Number 106, Carrollton, TX: Squadron/Signal Publications, 1990; David R. McLaren, *Black Widow: The Story of the Northrop P-61*, Colorado Springs: ViP Publishers, Inc., 1993; Garry R. Pape with John M. and Donna Campbell, *Northrop P-61 Black Widow: The Complete History and Combat Record*, Osceola, WI: Motorbooks International, 1991.

43. Don Dessert. Interview with the author June 23, 1999.

44. Jim Allen. Interview with the author. May 30, 2007.

45. Ibid.

Chapter 4

1. For more information on the battle for Noemfoor, as well as General MacArthur's movement across northern New Guinea in Operation Cartwheel, see Harry A. Gailey. *MacArthur's Victory: The War in New Guinea, 1943-1944*. New York: Presidio Press, *2004*.

2. Bill McConnell. Interview with the author. June 23, 1999.

3. Ibid.

4. Harold Young. Interview with the author. June 25, 1999.

5. Bob Fowble. Interview with the author. June 25, 1999.

6. Eddy Ballard. Interview with the author. June 24, 1999.

7. Bill McConnell. Interview with the author. June 23, 1999.

8. Harold Young. Interview with the author. June 25, 1999.

9. Ibid.

10. Dick Stewart. Interview with the author. June 25, 1999.

11. Don Dessert. Interview with the author. June 23, 1999.

12. Many conversations with Al Lukas.

13. Jesse Henry Gardner. *Beachheads and Black Widows: A South Pacific Diary*. Pittsburgh, PA: Hermmann Printing and Litho, Inc., 1995, p.69.

14. Ralph Dunavant. Interview with the author. June 24, 1999.

15. Charles Bailey. Letter to Eddie Barkhouse. March 27, 2998.

16. 419th NFS July 1944 operations are contained in 419th NFS. *Historical Record of 419th Night Fighter Squadron from 1 July through 31 July 1944.* Dated September 20, 1944.

17. Al Lukas flight log; conversations with Lukas; and 419th NFS. *Pilot's Individual Narrative Combat Report*. Dated August 6, 1944.

18. Eddie Barkhouse. Interview with the author. June 24, 1999.

19. 419th NFS August 1944 operations are contained in 419th NFS. *Historical Record of 419th Night Fighter Squadron from 1 August through 31 August 1944.* Dated October 7, 1944.

20. Bob O'Brien. Interview with the author. June 23, 1999.

21. Charles Bailey. Letter to Eddie Barkhouse. March 27, 2998.

22. Bob O'Brien. Interview with the author. October 6, 1998.

23. 419th NFS. *Final Mission Report*. Dated September 2, 1944.

24. 419th NFS September 1944 operations are contained in 419th NFS. 419th NFS. *Historical Record of 419th Night Fighter Squadron from 1 September through 30 September 1944.* Dated October 15, 1944.

25. 419th NFS October 1944 operations are contained in 419th NFS. 419th NFS. *Historical Record of 419th Night Fighter Squadron from 1 October through 31 October 1944.* Dated November 5, 1944.

Chapter 5

1. Jim Allen. Interview with the author. June 23, 1999.

2. Ibid.

3. Bob O'Brien. Interview with the author. June 24, 1999.

4. 419th NFS. *Historical Record of 419th Night Fighter Squadron from 1 August through 31 August 1944.* Dated October 7, 1944.

5. James Girdis. Interview with the author. June 24, 1999.

6. 419th NFS. *Final Mission Report*. Dated August 31, 1944.

7. Jim Allen. Interview with the author. June 23, 1999.

8. Ibid.

9. Jim Allen. Interview with the author. June 23, 1999 and Dick Stewart. Interview with the author. June 23, 1999.
10. 419th NFS. *Final Mission Report*. Dated September 11, 1944.

11. 419th NFS. *Aircraft Accident Report*. September 13, 1944; Dick Stewart. Interview with the author. June 24, 1999.

12. John Radzieta. Interview with the author. November 29, 2000; Dick Stewart. Interview with the author. June 25, 1999.

13. John Radzieta. Interview with the author. November 29, 2000.

14. Mission reviews for the previous month of September can be found in 419th NFS. *Historical Record of 419th Night Fighter Squadron from 1 September through 30 September 1944.* Dated October 15, 1944.

15. Jesse Henry Gardner. *Beachheads and Black Widows: A South Pacific Diary*. Pittsburgh, PA: Hermmann Printing and Litho, Inc., 1995.

16. 419th NFS. *Statement of Crew Chief, Aircraft Accident YP-61-A-10, AAF No. 42-5597.* October 4, 1944; Eddie Ballard. Interview with the author. June 24, 1999.

17. 419th NFS. *Final Mission Report*. Dated October 10, 1944; Don Dessert. Interview with the author. June 23, 1999.

18. Dick Stewart. Interview with the author. June 23, 1999.

19. Mission reviews for the previous month of October can be found in 419th NFS. *Historical Record of 419th Night Fighter Squadron from 1 October through 31 October 1944.* Dated November 5, 1944.

20. 419th NFS. *Final Mission Report*. Dated November 9, 1944,

21. 419th NFS. *Pilot's Individual Narrative Combat Report*. Dated November 10, 1944.

22. Jesse Henry Gardner. *Beachheads and Black Widows: A South Pacific Diary*. Pittsburgh, PA: Hermmann Printing and Litho, Inc., 1995.

23. Mission reviews for the previous month of November can be found in 419th NFS. *Historical Record of 419th Night Fighter Squadron from 1 November through 30 November 1944.* Dated December 5, 1944.

24. 419th NFS. *Aircraft Accident Report*. December 19, 1944.

25. 419th NFS. Final Mission Report. Dated December 30, 1944; Dick Stewart. Interview with the author. June 24, 1999.

26. Jesse Henry Gardner. *Beachheads and Black Widows: A South Pacific Diary*. Pittsburgh, PA: Hermmann Printing and Litho, Inc., 1995.

27. Jim Allen. Interview with the author. June 23, 1999, and Dick Stewart. Interview with the author. June 23, 1999.

28. 550th NFS. *History of the 550th NFS 1 January 1945 – 28 February 1945*. March 7, 1945.

29. 550th NFS. *Strafing Mission*. February 25, 1945.

Chapter 6

1. For more information on the battle for Morotai, see United States Army Far East Command. *Outline of the Battle for Morotai*. Tokyo: U.S. Army Far East Command, 1946; Gary Null, *The U.S. Army Air Forces in World War II: Weapon of Denial : Air power and the Battle for New Guinea,* Washington, DC: Air Force History and Museums Program, 1995; and Robert R. Smith, *The Approach to the Philippines*, Washington, DC: Office of the Chief of Military History, 1953.

2. Don Dessert. Interview with the author. June 23, 1999.

3. Sandy Bauman. Interview with the author. June 24, 1999.

4. 419th NFS. *Final Mission Report*. Dated December 6, 1944.

5. Harold Young. Interview with the author. June 24, 1999; 419th NFS. *Aircraft Accident Report*. December 8, 1944.

6. 419th NFS. *Final Mission Report*. Dated December 10, 1944.

7. John Graham. Interview with the author. June 25, 1999.

8. 419th NFS. *Final Mission Report*. Dated December 17, 1944.

9. Clem Green. Interview with the author. June 7, 1998.

10. 419th NFS. *Final Mission Report*. Dated December 28, 1944.

11. December operations of the 419th NFS can be found in 419th NFS. *Historical Record of the 419th NFS from 1 December 1944 to 31 December 1944*. Dated January 16, 1945.

12. John Graham. Interview with the author. June 25, 1999; 419th NFS. *Final Mission Report*. Dated January 9, 1945.

13. Jesse Henry Gardner. *Beachheads and Black Widows: A South Pacific Diary*. Pittsburgh, PA: Hermmann Printing and Litho, Inc., 1995.

14. Clem Green. Interview with the author. June 7, 1998.

15. 419th NFS. *Final Mission Report*. Dated January 11, 1945.

16. Jesse Henry Gardner. *Beachheads and Black Widows: A South Pacific Diary*. Pittsburgh, PA: Hermmann Printing and Litho, Inc., 1995.

17. John Radzieta. Interview with the author. June 23, 1999.

18. 419th NFS. *Aircraft Accident, P-61-B, 42-39447, Preliminary Report*. January 13, 1945.

19. 550th NFS. *History of the 550th Night Fighter Squadron 1 December 1944 to 28 February 1945*. March 1945. The Mid-Atlantic Air Museum recovered the P-61 in 1991. See www.maam.org.

Chapter 7

1. Clifford Johnson. Interview with the author. March 19, 2004.

2. United States Army. *Reports of General MacArthur. Japanese Operations in the Southwest Pacific Area. Volume II - Part II.* Compiled from Japanese Demobilization Bureaux Records. Facsimile edition 1994. Held by the Center for Military History.

3. For an introduction to the Southern Philippines Campaign see *Southern Philippines. The U.S. Army Campaigns of World War II*. U.S. Army Center of Military History.

4. The Palawan massacre of American prisoners is depicted in the opening scenes of the 2005 Miramax film entitled *The Great Raid*. For a general history of the massacre, see Dennis Wrynn, "American Prisoners of War: The Palawan Massacre," *World War II* magazine, November 1997. See also Bob Wilbanks, *Last Man Out: Glenn McDole, USMC, Survivor of the Palawan Massacre in World War II*, Jefferson, NC: McFarland & Company, Inc., 2004.

5. 419th NFS. *Historical Record of 419th Night Fighter Squadron from 1 March through 31 March 1945.* Dated April 5, 1945.

6. Ibid.

7. Ibid.

8. Ibid.

9. Ibid.

10. Bob O'Brien. Interview with the author. June 24, 1999.

11. James Girdis. Interview with the author. June 23, 1999.

12. Bob Wright. Interview with the author. June 24, 1999.

13. Bob O'Brien. Interview with the author. June 24, 1999.

14. Charles Pomeroy. Interview with the author. June 16, 1998.

15. 419th NFS. *Historical Record of 419th Night Fighter Squadron from 1 April through 30 April 1945.* Dated May 5, 1945.

16. Ibid.

17. Ibid.

18. Bob O'Brien. Interview with the author. July 11, 1999.

19. 419th NFS. *Historical Record of 419th Night Fighter Squadron from 1 April through 30 April 1945.* Dated May 5, 1945.

20. Ibid.

21. Ibid.

22. 419th NFS. *Final Mission Report*. Dated June 14, 1945.

23. 419th NFS. *Historical Record of 419th Night Fighter Squadron from 1 June through 30 June 1945.* Dated July 4, 1945.

24. 419th NFS. *Historical Record of 419th Night Fighter Squadron from 1 July through 31 July 1945.* Dated August 5, 1945.

25. 550th NFS. *History of the 550th NFS, 1 March 1945 to 1 May 1945*. Dated May 7, 1945.

26. 550th NFS. *Final Mission Report*. March 25, 1945.

27. 550th NFS. *Final Mission Report*. March 30, 1945.

28. 550th NFS. *Final Mission Report*. April 2, 1945.

29. 550th NFS. *Final Mission Report*. April 3 1945.

30. 550th NFS. *Final Mission Report*. April 5 1945.

31. 550th NFS. *Historical Data*. June 10, 1945.

32. 550th NFS. *Historical Data*. July 10, 1945.

33. 550th NFS. *Historical Data*. August 16, 1945.

34. United States Army. *Reports of General MacArthur. Japanese Operations in the Southwest Pacific Area. Volume II - Part II*. Compiled from Japanese Demobilization Bureaux Records. Facsimile edition 1994. Held by the Center for Military History.

35. For an introduction to the Southern Philippines Campaign see *Southern Philippines. The U.S. Army Campaigns of World War II*. U.S. Army Center of Military History.

36. Jesse Henry Gardner. *Beachheads and Black Widows: A South Pacific Diary*. Pittsburgh, PA: Hermmann Printing and Litho, Inc., 1995.

37. Ibid.

38. 419th NFS. *Historical Record of 419th Night Fighter Squadron from 1 April through 30 April 1945*. Dated May 5, 1945.

39. Jesse Henry Gardner. *Beachheads and Black Widows: A South Pacific Diary*. Pittsburgh, PA: Hermmann Printing and Litho, Inc., 1995.

40. Ibid.

41. Ibid; 419th NFS. *Aircraft Accident Report*. Dated April 10, 1945.

42. Jesse Henry Gardner. *Beachheads and Black Widows: A South Pacific Diary*. Pittsburgh, PA: Hermmann Printing and Litho, Inc., 1995.

43. Don Dessert. Interview with the author. June 25, 1999.

44. 419th NFS. *Historical Record of 419th Night Fighter Squadron from 1 June through 30 June 1945*. Dated July 4, 1945.

45. 419th NFS. *Aircraft Accident Report*. Dated July 11, 1945.

46. 419th NFS. *Historical Record of 419th Night Fighter Squadron from 1 June through 30 June 1945*. Dated July 4, 1945.

47. Clem Green. Interview with the author. June 7, 1998.

48. 550th NFS. *Historical Data*. August 16, 1945.

Chapter 8

1. 419th NFS. *Historical Report of the 419th Night Fighter Squadron from 1 August through 31 August 1945*. September 5, 1945.

2. 550th NFS. *Historical Data*. September 16, 1945 and David R. McLaren. *Black Widow: The Story of the Northrop P-61*. Colorado Springs: ViP Publishers, 1993, pp. 75-76.

3. William N. Hess. *Pacific Sweep: The 5th and 13th Fighter Commands in World War II*. Garden City, NY: Doubleday and Company, Inc., 1974, p. 194.

4. For more information on the P-61C, see Larry Davis and Dave Menard. *P-61 Black Widow in Action*. Aircraft Number 106, Carrollton, TX: Squadron Signal Publications, 1990 and Warren E. Thompson. *Northrop P-61 Black Widow*. Warbird Tech Series volume 15, North Branch, MN: Specialty Press, 1997.

5. These figures are recorded in Kenn C. Rust and Dana Bell. *The 13th Air Force Story in World War II*. Temple City, CA: Historical Aviation Album, 1981. Anyone wanting to read more about the history of the 13th Air Force can consult Kenn C. Rust and Dana Bell, *13th Air Force Story In World War II*. Temple City, CA: Historical Aviation Album, 1981; William Wolf. *13th Fighter Command in World War II: Air Combat over Guadalcanal and the Solomons*. Atglen, PA: Schiffer Military History, 2004; William N. Hess, *Pacific Sweep: The 5th and 13th Fighter Commands in World War II*. Garden City, NY: Doubleday and Company, 1974.

6. D. Clayton James. The Years of MacArthur. Volume II, 1941-1945. Boston: Houghton Mifflin Company, 1975, pp. 496-497.

7. Kristen Alexander, "'Cleaning the Augean stables'. The Morotai Mutiny?" *Sabretache*. Military Historical Society of Australia, 2004.

Index